AF553195

WORLD GREAT MATHEMATICIANS

WORLD GREAT MATHEMATICIANS

Edited by

G.R. CHHATWAL

RAJ KUMAR S.C. ANAND

S.K. ARORA K. NARANG

ANMOL PUBLICATIONS PVT. LTD.

NEW DELHI - 110 002 (INDIA)

ANMOL PUBLICATIONS PVT. LTD.
Regd. Office: 4360/4, Ansari Road, Daryaganj,
New Delhi-110002 (India)
Tel.: 23278000, 23261597, 23286875, 23255577
Fax: 91-11-23280289
Email: anmolpub@gmail.com
Visit us at: www.anmolpublications.com

Branch Office: No. 1015, Ist Main Road, BSK IIIrd Stage
IIIrd Phase, IIIrd Block, Bangalore-560 085 (India)
Tel.: 080-41723429 • Fax: 080-26723604
Email: anmolpublicationsbangalore@gmail.com

World Great Mathematicians

First Edition 1993
Reprint, 1999, 2002, 2004, 2010
ISBN 81-7041-761-9

PRINTED IN INDIA

Printed at Mehra Offset Press, Delhi.

PREFACE

This encyclopaedia presents biographical entries on internationally known scientists — biologists, chemists, physicists and mathematicians, who have made important contributions to science. These scientists have been selected from earlier history to the 20th century. This also covers a selection of people from philosophy, engineering and technology. The entries contain biographical data place and date of birth, posts held etc. but do not give exhaustive personal details about the subject's family, prizes, honorary degrees; etc. However a sincere effort has been made to include their main scientific achievements and the nature and importance of these achievements. Further, an attempt is made to include people who have produced major advances in theory or have made influential or well known discoveries. Certain entries have been quite lengthy because we have found them interesting. It is hoped that the reader will find the entries useful.

This encyclopaedia will be of immense value to the students of biology, chemistry, mathematics, and physics and to research scientists and engineers working on science projects.

The working of this nature is essentially a team work that has been accomplished with a sense of challenge and devotion. Finally we are most grateful for the patience of all those whom we bothered in the search after assistance or information, whether friends, colleagues or contributors.

Editors

Abel, Niels Henrik. (b. Aug 5, 1802, Finnoy, Norway; d. Apr. 6, 1829; Froland, Norway). Norwegian mathematician. Abel, the son of a poor pastor, studied at the University of Christiana (Oslo). After the death of his father, Abel had to support a large family; he earned what he could by private teaching and was also helped out by his teacher. He was eventually given a grant by the Norwegian government to make a trip to France and Germany to visit mathematicians. In Germany he met the engineer and mathematician August Crelle, who was to be of great assistance to him. Crelle published Abel's work and exerted what influence he could to obtain him a post in Germany. Tragically Abel died just when Crelle had succeeded in getting him the chair in mathematics at Berlin. With Evariste Galois (whom he never met), Abel founded the theory of groups (commutative groups are known as *Abelian groups* in his honor), and his early death ranks as one of the great tragedies of 19th century mathematics. One of Abel's first achievements was to solve the longstanding problem of whether the general quintic (of the fifth degree) equation was solvable by algebraic methods. He showed that the general quintic is not solvable algebraically and sent this proof to Karl Gauss, but unfortunately Gauss threw it away unread, having assumed that it was yet another unsuccessful attempt to solve the quintic. Abel's greatest work was in the theory of elliptic and transcendental functions. Mathematicians had previously focused their attention on problems associated with elliptic integrals. Abel showed that these problems could be immensely simplified by considering the inverse functions of these integrals — the so-called 'elliptic functions'. He also proved a fundamental theorem, *Abel's theorem*, on transcendental functions, which he submitted to Augustin Cauchy (and unfortunately fared no better than with Gauss). The study of elliptic functions inaugurated by Abel was to occupy many of the best mathematicians for the remainder of the 19th century. He also made very important contributions to the theory of infinte series.

Adhemar, Alphonse Joseph. (b. February 1797; Paris; d. 1862; Paris). French mathematician. Adhemar was a private mathematics tutor who also produced a number of popular mathematical textbooks. His most important scientific work was his *Les Revolutions de la mer* (1842) in which he was the first to propose a plausible mechanism by which astronomical events could produce ice ages on Earth. It had been known for some time that while the Earth moved in an elliptical orbit around the Sun it also rotated about an axis that was tilted to its orbital plane. Because the orbit is elliptical and the Sun is at one focus, the Earth is closer to the Sun at certain times of year. As a result, the southern hemisphere has a slightly longer winter than its northern counterpart. Adhemar saw this as a possible cause of the great Antarctic icesheet for, as this received about 170 hours less solar radiation per year than the Arctic, this could just be sufficient to keep temperatures cold enough to permit the ice to build up. Adhemar was also aware that the Earth's axis does not always point in the same direction but itself moves around a small circular orbit every 26000 years. Thus he postulated a 26000-year cycle developing in the occurrence of glacial periods, but his views received little support.

Albert Einstein. Albert Einstein was a great genius son of the earth who left a permanent impact of his scientific work in the field of physics. For his contributions, he is known as the father of modern physics. Although he was born at Ulm in Germany, his family moved to Munich when he was one year old. He was very shy in his childhood. Since his mother was fond of playing piano, he also learnt playing it from her mother. Sometimes he was so absorbed in music that he did not care even for his meals. Right from the beginning Einstein was interested in science. Once his father gifted him a watch in which a magnetic compas was fitted in the chain. He forgot about the watch and started asking his father many questions about the magnetic compass. Einstein was very sharp in mathematics but just mediocre in other subjects. When he was 15, his family moved to Italy. From there he was sent to Switzerland. There he appeared for an examination to take admission in the Zurich University, but he failed. Next year he appeared again after full preparations and cleared the examination for admission in the university. In 1900, he completed his education, and became a citizen of Switzerland. He wanted to become a teacher, but could not get a teaching job of his choice. Finally, he jointed the Swiss Patent Office as a clerk. During these days young Einstein fell in love with a Yugoslav science student, Mileva Marec, and married her in 1903. The couple had two sons. Einstein used to take his child in a pram for a walk. During his walk he always carried a notebook with him.

His mind was always engaged in thinking something about science. He would often stop the pram and write something in his notebook regarding problems of mathematics and physics. Later, when the notebook was scrutinized, it contained many solutions regarding the problems of the universe. In 1905, at the age of 26, Einstein got his Ph.D. degree from Zurich University. In the same year, he published five research papers. Almost overnight, Einstein became a world famous scientist. In one paper, he showed that when light falls on metals like potassium, tungston, etc., they emit electrons. He called these electrons 'photoelectrons' and the effect 'photoelectric effect'. For propounding the theory of this effect, he was awarded Nobel Prize in 1921. In his second paper, he gave a theory for Brownian Motion, according to which the motion of free particles in a liquid is due to collision of particles with the molecules of the liquid. His third paper was related to the "Special Theory of Relativity" in which he showed that the physical quantities like mass, length and time are not constant, but vary with the velocity of the body. This paper stunned the world. This theory was so complicated that only few scientist of the world could understand it fully at that time. In his fourth paper, he gave the revolutionary idea establishing the equivalence of matter and energy. According to this theory, if one pound of matter is converted into energy then it will give an energy equal to that obtained by burning seven million tons of dynamite. The atomic bomb was the result of this equation. In his fifth paper, Einstein proved that light travels in the form of particles called photons. In 1916, he published a paper on the 'General Theory of Relativity', and explained the way the force of gravity works. In 1933, Germany was in the grip of brutality and dictatorship of Hitler.

The condition of Jew was very miserable. For this, Einstein started opposing Hitler. His activities made Hitler very angry. During this period Einstein was invited to give a lecture in America. While he was in America his friends wrote to him not to come back to Germany where he was likely to be punished. Einstein accepted the advice of his friends and decided to stay on in America where he was accorded a senior position in Princeton University. He worked on the development of Atom Bomb in the regime of President Roosevelt. On seeing the terrible effects of the atom bomb on Hiroshima and Nagashaki, Einstein became very sad and worked throughout his life on the peaceful uses of atomic energy. After retiring from Princeton University, in 1845, Einstein continued to serve science. On April 18, 1955, this great scientist died in a Princeton hospital. His end came while he was fast asleep. Einstein was not only a scientist, but also a great apostle of peace. To honour Einstein, an element 'Einsteinium' has been named after him. After his death, his brain was taken out and preserved in Princeton hospital. A number of scientist has studied it to know more about the mystery of the genius of this great scientist.

Aleksandrov, Pavel Sergeevich. (b. Bogorodska, Russia, 7.5.1896; d. Moscow, USSR, 16.11.1982). Russian mathematician who contributed primarily to the development of topology, a modern form of geometry. While a student at the University of Moscow, he became interested in abstract set theory, as founded by Cantor. In the early 1920s he vigorously developed set-theoretic topology with his friend P.S. Uryson. They developed the theory of general topological spaces and compact topological spaces, initiated by Frechet and F. Hausdorff.

After Uryson died in an accident in 1924. Aleksandrov turned increasingly towards investigating combinatorial or algebraic topology, as founded by POINCARE. He thus created a homological theory of dimension. With the Swiss mathematician H. Hopf he planned a three-volume comprehensive treatise on topology. Only the first volume *Topologie I* (Berlin, 1935) was published, but this has been highly influential. A founder of the Moscow school of topology, he trained many Russian mathematicians in research and wrote extensively, both research works and student textbooks.

Alfred North Whitehead. Alfred North Whitehead was born in 1861 at Ramsgate in the Isle of Thanet, Kent, England. His father, like his grandfather, worked as the headmaster of a local private school although Alfred's father resigned his position in 1860 to become ordained as a clergyman of the Anglican church. After an appointment as Vicar of St. Peter's Parish, Whitehead's father "became influential among the clergy of East Kent, occupying the offices of Rural Dean, Honorary Canon of Canterbury, and Proctor in Convocation of the Diocese." Whitehead attributed his father's influence to his popularity with thee parish members and his daily visits to the parish's thre parochial schools. At the age of fourteen Albert was sent to school at Sherborne in Dorsetshire in southern England where he studied Greek and Latin interspersed with mathematics and history. He played cricket and soccer and developed an interest in the poetry of Wordsworth and Shelley. Whitehead entered Trinity College, Cambridge, in 1880. The lectures he attended during his undergraduate career dealt only with pure and applied mathematics. He was awarded a fellowship at Trinity in 1885; he was later appointed to the position of senior lecturer though he gave it up when he moved to London in 1910. While teaching at Cambridge, Whitehead published his first book, *A Treatise on Universal Algebra*, which developed the geometrical ideas originally espoused by the talented but obscure German mathematician Hermann Grassmann in his book, *Ausdehnungslehre*. "*Universal Algebra* deals with the theory of matrices and other noncommutative algebras, also with non-Euclidean geometry, but its major concern is the algebra of symbolic logic." Its examination of matrix theory proved to be especially useful to Heisenberg and Born in their development of their system of quantum mechanics. Whitehead's book received much attention and led to his election to the Royal Society in 1903. During his final years at Cambridge, Whitehead began an extensive collaboration with Bertrand Russell, a former student, on their massive work, *Principia Mathematica*, which attempted to "bring order to the sadly disarrayed study of the foundations of mathematics." Although *Principia* was only partially successful in proving that mathematics is a part of logic, it has been acknowledged to be the single greatest contribution to logic since Aristotle and one of the most remarkable modern collaborations in any discipline. After moving to London, Whitehead published his *Introduction to Mathematics*, a popular treatment of the subject, then joined the faculty of University College, London. In 1914 he became a professor at the Imperial College of Science and Technology in Kensington where he remained for ten years until accepting an offer to join the philosophy department at Harvard University. Whitehead stayed at Harvard until his retirement in 1937, ten years before his death at the age of eighty-six. Soon after joining the Harvard faculty,

Whitehead gave a series of lectures (from which the present selection is taken) which were published in 1925 under the title *Science and the Modern World.* This collection examined the effects of scientific developments on three centuries of Western culture. It was to be perhaps the most influential of Whitehead's writings, enjoying great popularity with both academic and general audiences. It was in this book that Whitehead began to develop in detail his philosophy of science: His program in regard to natural philosophy, or philosophy of science was to deduce scientific concepts from the simplest elements of our perceptual knowledge. Put otherwise, the object of the analysis is nature, and by nature Whitehead means the world, as presented to our awareness.... The ultimate fact for sense-awareness is an event. Events are the most concrete facts of nature. They pass and are gone.... In contrast to events, objects recur. This means that objects characterize events. Objects are abstractions and enter experience through intellectual recognition.... Objects are, however, never merely in a single place at a single time. Each object has a complex ingression so that it is in some sense ingredient throughout nature.... Events recur, and the objects relate to each other, in a four-dimensional spacetime manifold, termed the extensive continuum. In 1927 Whitehead gave a series of lectures at the University of Edinburgh which were published two years later as *Process and Reality.* This book laid out Whitehead's murky philosophy of the organism; a unique cosmological blending of rationalism and religion. Whitehead argued in a throwback to the Greek philosopher Heraclitus that the Universe is "one of process, of becoming and perishing." Yet Whitehead's metaphysical approach did not escape criticism: "Its animism seems to reintroduce a brand of pohocus-cus which rational thought has long sought to extirpate: its mathematical elements are forbidding and its romantic strains confusing. Moreover, it is couched in an involved idiom made more involved by the strange, unlovely words—concresence, prehension, appetition, etc.—that Whitehead coined. Yet even those who are most strongly repelled by the philosophy of the organism recognize in its formulation the many profound observations, the honesty and daring imagination of a truly wise human being."

Al-Khwarizmi, Abu Ja'far Muhammad Ibn Musa. (b. c. 800;Khwarizm, now Khiva in the Soviet Union; d. c. 847) .Arab mathematician, astronomer, and geographer. Al-Khwarizmi's importance lies chiefly in the knowledge he transmitted to others. Very little is known about his life except that he was a member of the academy of sciences in Baghdad, which flourished during the rule (813-33) of caliph al-Ma'mun. Al-Khwarizmi's main astronomical treatise and his chief mathematical work, the *Algebra,* are dedicated to the caliph. The *Algebra* enlarged upon the work of Diophantus and is largely concerned with methods for solving practical computational problems rather than algebra as the term is now understood. Insofar as he did discuss algebra, al-Khwarizmi confined his discussion to equations of the first and second degrees. His astronomical work, *Zij al-sindhind,* is also based largely on the work of other scientists. As with the *Algebra*, its chief interest is as the earliest Arab work on the subject still in existence. Al-Khwarizmi's other main surviving works are a treatise on the Hindu system of numerals and a treatise on geography. The Hindu number system, with its epoch-making innovations, for example the incorporation of a symbol for zero, was introduced to Europe via a Latin translation (*De numero indorum;* On the Hindu Art of Reckoning) of al-Khwarizmi's work. Only the Latin translation remains but it seems certain that al-Khwarizmi was the first Arab mathematician to expound the new number system systematically. The term 'algorithm' (a rule of calculation) is a corrupted form of his name. His geographical treatise marked a considerable improvement over earlier work notably in correcting some of the influential errors and misconceptions that had gained currency owing to Ptolemy's *Geography.*

Amici, Glovanni Battista. (b. Mar. 25, 1786; Modena, Italy; d. Apr. 10, 1864, Florence, Italy). Italian mathematician and Instrument maker. Amici was professor of mathematics at the Univeristy of Modena and in 1835 became the director of the observatory at the Royal Museum in Florence. He made great improvements in the design of parabolic mirrors for reflecting telescopes, and constructed and designed prismatic spectroscopes. In 1840 he made two achromatic objective lenses with diameters of 9.5 and 11 inches (24 and 28 cm), which were used by Giovanni Donati. He also made advances in microseopy, improving the compound microscope and using it to study plant reproduction.

Apollonius. (b. c. 262 BC; Perga now in Turkey; d. c. 196 BC Alexandria, Egypt). Greek mathematician and geometer. Apollonius studied in Alexandria possibly under pupils of Euclid, and later he taught there himself. One of the great Greek geometers, Apollonius's major work was in the study of conic sections and the only one of his many works to have survived is his eight-book work on this subject, the *Conics.* Apollonius's work on conics makes full use of the work of his predecessors, notably Euclid and Conon of Samos, but it is a great advance in terms of its thoroughness and systematic treatment. The *Conics* also contains a large number of important new theorems that are entirely Apollonius's creation. He was the first to define the parabola, hyperbola, and ellipse. In addition, he considered the general problem of finding normals from a given point to a given curve (i.e. lines at right angles to a tangent at a point on the curve). Apart from the geometrical work that has survived, Apollonius is known to have contributed to optics — in particular to the study of the properties of mirrors of various shapes. This work, however, is now lost.

Archimedes. At the beginning of this century, in 1901 to be exact, the Greek scholar Papadopoulos Kerameus, in the course of search among the ancient volumes and manuscripts in the library of the Monastery of the Holy Sepulchre in Jerusalem, came across a very old parchment manuscript. The document was falling to pieces and its text was almost illegible, but Kerameus examined it with great care and discovered that it dealt with mathematics and was clearly of very great historical interest. When attempts to decipher it proved of no avail, the manuscript was sent to Constantinople, where J.L. Heiberg, a Danish historian and expert on Greek antiquities, was invited to work on it. After a great deal of trouble, he managed to clarify the text and figures, and announced to the world of scholarship a most exciting piece of news. The manuscript which had come to light was a short but important treatise by Archimedes, hitherto unknown, called *The Method.* It was addressed to his friend Eratosthenes and explained how he arrived at his conclusions in his research on the theory of areas and volumes. More than 2,000 years had passed since the death of Archimedes, and the discovery of one of his works (though not one of the most important) was a great event for science and aroused interest throughout the whole world. This is an indication of how great is the reputation of Archimedes of Syracuse. Archimedes was born at Syracuse in Sicily in the year 287 B C. He grew up in an atmosphere of science and learning, for his father Pheidias was an astronomer. From an early age Archimedes proved to be an eager student of quite unusual intelligence. He travelled in Egypt, and it was probably in Alexandria that he met and became friendly with the famous Eratosthenes of Cyrene, the philosopher who made an estimate of the circumference of the earth. It seems likely that it was the association with Eratosthenes, as well as his family tradition, that led Archimedes to take an interest in astronomy. On returning to Syracuse he applied himself to various studies — mathematics, physics, mechanics, astronomy — and became pre-eminent in all of them. Even today, with the help of teachers and explanatory text-books, students find these subjects difficult enough. The ancient philosophers worked from first principles, without any such assistance. In 216 B C, when Archimedes was over 70 years old, Hieron, the king of Syracuse, died; it is believed that Archimedes was related to him. It was the time of the Second Punic War, and Syracuse had decided to enter into an alliance with the Carthaginians. In consequence of this the Romans sent an army, under the command of the Consul Claudius Marcellus, to besiege the town. Archimedes was old and would have preferred to pursue his studies in peace and quiet; but his fellow-citizens, who knew well his wisdom and ingenuity, turned to him for help in the defence of the city. Archimedes accepted the invitation rather reluctantly, and the Romans soon learned of his ability as an inventor and engineer. The crew of a Roman ship, which had ventured right under the enemy fortifications, saw what appeared to be a huge pair of pincers appear from the walls, grasp the hull of the ship between its jaws, shake it and almost demolish it. The pincers were a war machine invented by Archimedes; it worked by means of levers and pulleys, a type of mechanism on which the old scientist was a great expert. At the same time giant catapults hurled showers of heavy spears and stones on the ships at anchor some distance from the walls. They broke the decks and sides of the ships, tore down the masts, mowed down the crews. Great rocks were also thrown on the Roman ships from the highest part of the walls. There is also a story that Archimedes set fire to some ships by directing the sun's rays onto them from great mirrors, but this is hardly credible and is probably only a legend. While the Roman fleet was kept at a distance in this way, teams of workers and slaves toiled ceaselessly in the deep caves and old stone quarries of the town to build further formidable machines planned by the old scientist. Thus it was that Archimedes, at the age of 75, met his death. He was one of the greatest philosophers of ancient times. He loved science and learning for its own sake. He brought the study of geometry up to a point where it foreshadowed integral calculus, one of the most important principles of present-day higher mathematics. His contribution to the defence of Syracuse clearly illustrates his remarkable mastery of the practical applications of science. However, he regarded these achievements as of small importance and made practically no reference to them in his own writings. The great mathematicians of the seventeenth century - Pascal, Fermat, Huygens and Newton — made extensive use of the discoveries and methods of Archimedes, as did the founders of modern physics, Kepler, Galileo and Torricelli. These men studied his works and used his methods and principles in their research, regarding him as pupils regard a learned and respected master. One of the feats of engineering attributed to Archimedes was the launching of a large, fully laden, three-masted ship by means of an apparatus of levers and pulleys. The watching crowd were amazed and believed that he was using magical powers. He also invented the ingenious 'water-screw', a device for raising water to a higher level, and he worked on the properties of light, using mirrors. The most famous of all his achievements was to establish the so-called *Archimedes Principle,* which states that a body immersed in liquid displaces its own volume of the liquid, irrespective of its weight. Our modern tradition of scientific thinking makes this seem obvious, but it was far from being so in Archimedes' day. The story goes that King Hieron gave Archimedes a golden crown, which he suspected of being alloyed with silver, and told him to test this without damaging it. Since gold is much heavier than silver, a given weight of it will, if immersed in water, displace less volume than the same weight of silver the same weight of an alloy of the two will displace an intermediate amount of water. It is said that the realisation of this principle, which would solve his problem, came suddenly to Archimedes when he was lying in his bath. He noticed how when he got in, his body displaced the water, making the level in the bath rise. He is supposed, then, to have jumped out and run down the street, unclothed, shouting 'Eureka! Eureka!' (I have found it). The first part of the story is very likely true but it is more probable that he solved the problem by hard thinking and methodical experiments. Thanks to their preseverance, the Romans finally captured Syracuse in 212 B C. After a heroic defence lasting four years, the Roman troops succeeded in entering and sacking the town on the day of the feast in honour of the goddess Artemis. During the ensuing disorder a Roman soldier entered the house of Archimedes, who was so absorbed in some geometrical calculations that he did not hear the shouting, the clash of arms and the clatter of horses' hoofs outside. The soldier looked suspiciously at the calm old man, who did not even turn at his entrance. He took two steps forward. Archimedes had drawn some diagrams on the ground, and at last became aware of two feet, shod in military sandals, threatening to tread on them, so he said in Latin (the soldier's language): 'Noli turbare circulos meos' (Please do not spoil my circles). Another version of his last words is: 'Noli, obsecro, istud disturbare' (I pray you, do not disturb that). What went on in the soldier's mind at that moment, no one can say. It is only known that he raised his sword and killed old Archimedes.

Ampere, André Marie. (b. Jan. 22, 1775; Lyons, France; d. June 10, 1836; Marseille, France). French physicist and mathematician. Ampere, the son of a wealthy merchant, was privately tutored, and to a large extent self-taught. His genius was evident at an early age. He was particularly proficient at mathematics and, following his marriage in 1799 he was able to make a modest living as a mathematics teacher in Lyons. In 1802 he moved first to Bourg-en-Bresse to take up an appointment, then to Paris as professor of physics and chemistry at the Ecole Centrale. His first publication was on the statistics of games of chance *Considérations sur la théorie mathématique du jeu* (1802; Considerations on the MathematicalTheory of Games) and his work at Bourg led to his appointment as professor of mathematics at the Lyceum of Lyons, and then in 1809 as professor of analysis at the Ecole Polytechnique in Paris. His talents were recognized by Napoleon, who in 1808 appointed him inspector general of the newly formed university system — a post Ampere held until his death. Ampere's most famous scientific work was in establishing a mathematical basis for electromagnetism. The Danish physicist Hans Christian Oersted had made the important discovery that the current passing through a wire could cause the movement of a magnetic compass needle. Ampere witnessed a demonstration of electromagnetism by Francois Arago at the Academy of Science on 11 September, 1820. He set to work immediately on his own investigations, and within seven days was able to report the results of his experiments. In a succession of presentations to the academy in the next four months, he developed a mathematical theory to explain the interaction between electricity and magnetism, to which he gave the name 'electrodynamics' (now more commonly: electromagnetism) to distinguish it from the study of stationary electric forces, which he christened 'electrostaties'. Having recognized the electric currents in wires caused the motion of magnets, and that a magnet can affect another magnet, he looked for evidence that electric currents could influence other electric cuı rents. The simplest example of this interaction is found by arranging for currents to flow through two parallel wires. Ampere discovered that if the currents passed in the same direction the wires were attracted to each other, but if they passed in opposite directions the wires were repelled. From this he went on to consider more complex configurations of loops, helices, and other geometrical figures, and was able to provide a mathematical analysis that allowed quantitative predictions. In 1825 he had been able to deduce an empirical law of forces (*Ampere's law*) between two current-carrying elements, which showed an inverse-square law (the force decreases as the square of the distance between the two elements, and is proportional to the product of the two currents). By 1827 he was able to give a precise mathematical formulation of the law, and it was in this year that his most famous work *Mémoire sur la théorie mathématique des phénoménes électrodynamiques uniquement déduite de l'expérience* (Notes on the Mathematical Theory of Electrodynamic Phenomena, Solely Deduced from Experiment) was published. Besides explaining the macroseopic effects of electromagnetism, he attempted to construct a microscopic theory that would fit the phenomenon, and postulated an electrodynamic molecule in which electric-fluid currents circulated, giving each molecule a magnetic field. In his honor, the unit of electric current is named for him, and in fact the ampere is defined in terms of the force between two parallel current-carrying wires.

Aristotle. No man in the history of human culture has had a deeper authority or a longer-lasting influence than Aristotle. His logic was the world's logic for nearly two millennia; his ideas on natural science endured throughout the Middle Ages and well into the Renaissance. It was only some nineteen centuries after his death that Aristotelian physics inspired so powerful a reaction against itself that it gave great impetus to the development of Galilean and, eventually, Newtonian physics. Aristotle was born in 384 B.C. in the northern Greek town of Stagira, which is why he is often referred to as the Stagirite. His father was court physician. At the age of seventeen Aristotle went to study at Plato's Academy in Athens, where the curriculum included not only mathematics and astronomy but also biology and medicine. After the death of Plato, Aristotle taught at the court of Hermias of Atarneus. In 343 B.C. he joined the court of Philip II of Macedon, where his principal task was the tutoring of Alexander the Great. Eight years later he founded his own school in Athens, known as the Lyceum. When Alexander died in 323 B.C. many Athenians turned against his supporters; Aristotle was forced to abandon Athens for Chalcis, where he died in 322 B.C. Aristotle, was less concerned than Plato about the relationship between the ideal world and the real world. One of his greatest contributions to science may have been his efforts to systematically classify plants and animals. He also sought to discover the primary forces of nature. According to Aristotle, the Universe is balanced by the opposite movements of the four elements. Air and fire move upward naturally while earth and water move downward, thereby preserving a general equilibrium. Aristotle used a fifth element, ether, to explain the movements of stars, which he supposed consisted of ether. His concept of potentiality or capacity was applied to much of his philosophy, especially the ideal of infinity. His potential infinity could be endlessly divided or reduced. But Aristotle was practical enough to understand the futility of attempting to endlessly divide an infinte series. Yet the distinction between potential and actual infinities enabled him to sidestep many of the paradoxes posed by Zeno. Aristotle also attempted to describe the essence of the human soul and the structuring forces of nature in terms of a unified entity. Despite his preoccupation with the organization of physical events, however, the material world often occupied less of Aristotle's attention that the more mysterious spiritual world. Aristotle's *Physics* founded the academic subject of physics. In it Aristotle provided a detailed system of speaking and reasoning about all facets of human experience which distinguished systematically between the processes of deduction and induction as well as among the theoretical, practical, and productive divisions of science. The system known as Aristotelian logic that was distilled from this work is a small part of his complex thought on logic, the part that he felt was appropriate for timeless propositions like those of mathematics. Early in the *Physics* he made it clear that a more dialectical logic was needed to deal with change because during a change from A to not-A, neither A nor not-A can hold. Aristotle argued from this that there must be more than two principles in physics because there could otherwise be no change from one to the other. Although a similar pattern of change is evidenced in modern quantum theory, there was no place for Aristotle's dialectics in science before Planck formulated his theory. Aristotle was also very careful in defining the line or one-dimensional continuum and noted that it is a totality of points that has a principle of connection. Today we speak not of a principle of connection but of a topology. We distinguish when necessary between the line as the set of its points, without connections, and the line as a topological space, with connections. The topological space may be expressed, for example, not as a set of points but as a set of intervals or neighborhoods of points. This scheme did not satisfy Aristotle at first because each interval has the same problematical infinity of points as the whole line and it is not stated how such intervals are to be specified. Descartes solved this problem by identifying the line with the totality of numbers:

positive, zero, negative, rational, and irrational. After Descartes an interval is defined as a set of numbers determined by two numbers A and B, consisting of all the numbers greater than A and less than B. We define the topological structure of the line by such intervals. Although Euclid had already laid down the axioms of his geometry, Aristotle was reluctant to use a mathematical model to describe his Universe. Aristotle's God did not model the Universe using basic geometrical shapes as Plato had supposed or numbers as Pythagoras had imagined; Aristotle presumed that the cosmos was governed by four kinds of cause: formal, material, efficient, and final.

Artin Emil. (b. Vienna, Austria-Hungary [now Austria], 3.3.1898; d. Hamburg, West Germany, 20.12.1962). Austrian mathematician. The analogies Artin detected between algebraic and analytic number theory have inspired much subsequent work, including the resolution of conjectures later made by André Weil partly on the basis of Artin's work. In his Leipzig thesis (1921) Artin studied quadratic extensions of the field of rational functions in one variable over a finite field, introduced a zeta function and proposed a hypothesis for it analogous to the Riemann hypothesis in the complex case. He established the hypothesis in various cases: in 1934 Hasse gave a more general proof, and in 1948 Weil gave a complete proof. In 1923, when he was a lecturer at the University of Hamburg, before holding the chair in mathematics (1926-37), Artin introduced his L-series, which generalize Dirichlet's, and using them produced his important general reciprocity theorem, a cornerstone of Abelian classified theory. Artin had considerable influence on the development of abstract algebra and on the applications of group theory to topology. He spent the period 1937-58 in the USA, from 1946 at Princeton University, where his influence is considerable. He returned to the University of Hamburg in 1958. Artin was an extremely talented amateur musician, playing the flute, clavichord and harpsichord. S. Lang & J.T. (eds), *The Collected Papers of Emil Artin* (Reading, Mass., 1965).

Atiyah, Michael Francis. (b. London, UK, 22.4.1929). British mathematician. A student of W.D.V. Hodge at Cambridge, Atiyah has made crucial contributions to the study of algebraic geometry by topological means, and then in the late 1950s with Hirzebruch he developed a tool, K-theory, which he has since put to great use in the theory of elliptic partial differential equations. His work in this area represents a major advance in the study of a topic which, at its simplest, includes the steady-state equation for heat distribution. His studies provide a topological meaning for terms previously defined analytically, a considerable conceptual gain. More recently he has investigated a geometric theory of instantons of interest to particle physicists. An enthusiastic collaborator with others, and a forceful popularizer of and propagandizer for mathematics, especially geometry, he exerts a beneficial influence on many branches of the subject and at many levels. He was awarded a Fields medal in 1966. Savilian professor of mathematics at Oxford (1963-9), he held the chair in mathematics at the Institute of Advanced Studies, Princeton (1969-72), before returning to St Catherine's College, Oxford, in 1973 as Royal Society research professor.

B

Babbage, Charles. (b. London, UK, 26.12.1791; d. London, 18.10.1871). British computer pioneer, born a century too soon. While they were undergraduates at Cambridge, Babbage, John Herschel and others founded the Analytical Society to introduce continental notation in calculus to England. From their work developed the modern school of English mathematics. Babbage's own early work was in the theory of functions and modern algebra, of which he was one of the main pioneers. Between 1822 and 1833 he developed a Difference Engine to form and print mathematical tables for navigation, etc. In 1830 he wrote *On the Decline of Science* and led a campaign to turn the Royal Society into a professional body. This the campaign failed to do, but it did result in the foundation of the British Association for the Advancement of Science. In 1832 he published *On the Economy of Machinery and Manufactures* (London), which placed the factory at the centre of political economy and exerted a strong influence on John Stuart Mill and Marx. Babbage twice stood for the newly reformed Parliament and in 1833 published *A Word to the Wise,* advocating life peerages. In 1837 he published *The Ninth Bridgewater Treatise,* which presented the deity as computer programmer. From 1834 to 1848 he developed many plans for 'analytical engines', which were versatile programmable mechanical calculators with a decimal number representation, and between 1848 and 1856 he considered the question of how to construct such. Technically it was practicable but it proved impossible to raise the necessary finance. In 1856 he returned to the analytical engines as a hobby for his old age. Babbage's engines incorporate an extraordinary range of ideas utilized in modern computing, including: program control, microprogramming, a range of peripherals, seperate store and mill, multiprocessing, and even array processing.

Babbage, Charles. (b. Dec. 26, 1792; Teignmouth, England; d. Oct. 18, 1871; London). British mathematician. Babbage, whose father was a banker, studied at Cambridge and played a major role in ending the isolationist attitudes prevalent in British mathematical circles in the early 19th century. In 1815 he helped to found the Analytical Society, which aimed to make the work of Continental mathematicians better known in Britain. Babbage's interest in stimulating British scientific activity was by no means confined to mathematics. In 1820 he was a founder of the Royal Astronomical Society and in 1834 of the Statistical Society, and he continued to attack the British public for their lack of interest in science. Among his inventions were a speedometer, and the locomotive 'cowcatcher'. Babbage also did mathematical work that contributed to the setting up of the British postal system in 1840. From 1828 to 1839 he was Lucasian Professor of Mathematics at Cambridge University. Babbage is best known for his pioneering work in designing and building a mechanical computer. Earlier mathematicians such as Blaise Pascal and Gottfried Leibniz had designed primitive computing machines, but Babbage's 'analytical engine', as he called it, was very much more sophisticated than anything that had been thought of before. Many key ideas subsequently taken up when modern electronic computers were being developed were due to Babbage; for example, the ideas of giving the machine instructions in the form of punched cards, having the results in the form of a print-out, and the need for the computer to have a 'memory' — all originate with Babbage. However although Babbage had these theoretical insights into computer design he was never able to complete the prototype of his analytical engine owing to the limitations of the purely mechanical technology available to him and to lack of money. Originally he had been able to get financial backing from the British government for the project, but as time went by without Babbage's completing the machine, the government refused to put any more money into the venture. Babbage then began to spend more and more of his own money on the machine. One keen supporter of the project who came to his aid was Lord Byron's daughter Ada, Countess of Lovelace, who had considerable mathematical ability. Lady Lovelace and Babbage devised a scheme aimed at winning enormous sums on horse races but unfortunately this was not a success and Babbage's financial situation continued to deteriorate. It was largely through Lady Lovelace's enthusiastic interest in Babbage's machine and her attempts to interest others in it that it is as well known as it is. Eventually Babbage used up all his own money on the computer project and the prototype remained incomplete. It still exists and can be seen in the Science Museum in London.

Balmer, Johann Jakob. (b. May 1, 1825; Lausanne, Switzerland; d. Mar. 12, 1898; Basel, Switzerland). Swiss mathematician. Balmer was not a professional scientist but worked as a school teacher in Basel from 1859. In 1885 he discovered that there was a simple mathematical formula that gave the wavelengths of the spectral lines of hydrogen — the *Balmer series*. This formula proved to be of great importance in atomic spectroscopy and in developing the atomic theory. Balmer arrived at his result purely from empirical evidence and was unable to explain why it yielded correct answers. Not until the further development of the atomic theory by Niels Bohr (q.v.) and others was this possible.

Bar-Hillel, Yehoshua. (b. Vienna, Austria-Hungary [now Austria], 8.9.1915; d. Jerusalem, Israel, 25.9.1975). Austria/Israeli logician, philosopher and theoretical linguist. From 1961 he was professor of logic and the philosophy of science at the Hebrew University, Jerusalem. He received his doctorate at the Hebrew University in 1947. then spent three years at Chicago University under Carnap, followed by a further three years at MIT (1950-3). Carnap's influence was decisive, confirming him in a rejection of nonrigorous approaches to philosophy. His career spanned a wide range of formal, applied and philosophical perspectives on language. He participated in the machine translation (MT) and cybernetics boom of the early 1950s, but turned increasingly to algebraic and mathematical linguistics and, under Chomsky's influence, to the application of logic to the study of natural language. A major contribution was to foster mutual awareness between linguists and logicians. He made technical contributions to formal syntax, and his work is characterized by an insistence on methodological consistency. In the late 1960s and early 1970s he encouraged the growth of pragmatics (a branch of linguistics concerned with relevance to speaker and context). The collection of essays *Language and Information* (Cambridge, Mass., 1964) deals especially with information theory and mathematical linguistics, and *Aspects of Language* (Jerusalem, 1970) shows his later interest in general issues in the philosophy of language and methodology of linguistics. He was a prominent figure in Israeli academic life. A Kasher (ed.), *Language in Focus: Foundations, Methods and Systems: Essays in Memory of yehoshua Bar-Hillel* (Dordrechi, 1976).

Barrow, Isaac. (b. 1630; London; d. May 4, 1677; London). British mathematician. Barrow obtained his degree in 1648 but because of the unrest of the Cromwellian period was unable to take up an academic post until 1660 when he became professor of Greek at Cambridge University. He held the Lucasian Chair in Mathematics at Cambridge University from 1664 until 1669 when he resigned the chair in favor of his brilliant pupil Isaac Newton. Barrow's own mathematical work came very close to anticipating some of the basic concepts of the differential calculus; in his *Lectiones geometricae* (1670) he describes a method of calculating tangents to curves. In addition to being a mathematician Barrow was also a noted classicist and theologian and he produced a widely used translation of Euclid's *Elements*.

Bartholin, Erasmus. (b. Aug. 13, 1625; Roskilde, Denmark; d. Nov. 4, 1698; Copenhagen). Danish mathematician. Bartholin, the son of Caspar and brother of Thomas Bartholin, who were both distinguished anatomists, was educated in Leiden and Padua, where he obtained his MD in 1654. After further travel in France and England he returned to Denmark in 1656 and held chairs in mathematics and medicine at the University of Copenhagen from 1657 until his death. Bartholin worked on the theory of equations and with Olaus Romer made an unsuccessful attempt to calculate the orbits of the comets prominent in the late 1660s. He is however best remembered for his discovery of double refraction announced in his *Experimenta crystalli Islandici disdiaclastici* (1669). In it he described how Icelandic feldspar (calcite) produces a double image of objects observed through it. This discovery greatly puzzled scientists and was much discussed by Newton and Christian Huygens, who tried unsuccessfully to incorporate the strange phenomenon into their respective theories of light. Double refraction proved remarkably recalcitrant to all proposed explanations for well over a century and it was only with the work of Etienne Malus on polarized light in 1808, and that of Augustin Fresnel in 1817, that Bartholin's observations could at last be understood.

Bartlett, Maurice Stevenson. (b. Scrooby, Notts., UK, 18.6.1910). British statistician, with ICI from 1934 to 1938 and since then holding academic posts in mathematics and statistics at Cambridge (1937-47), Manchester (1947-60), London (1960-7) and Oxford (1967-75) where he is now emeritus professor of biomathematics. Starting from the problem of establishing the basic virulence of epidemics, Bartlett built up a general methodology of inference for 'point-processes'. He saw, as had many others, that it is relatively easy to model the course of an epidemic in a known population, if the underlying infectivity is known. It is very difficult, however, to infer the basic virulence from the number of actual cases since the same disease may be either exploding or dying out. By taking as his data the time intervals between cases, Bartlett succeeded in applying reliable statistical analysis (see *An Introduction to Stochastic Processes.* Cambridge, 1955, ^{3}1978).

Benjamin Peirce. Oliver Wendell Holmes' memorial poem on one of the greatest of the American mathematicians begins: To him the wandering stars revealed
The secrets in their cradle sealed;
The far-off, frozen sphere that swings
Through ether, zoned with lucid rings;

The orb that rolls in dim eclipse.
Wide wheeling round its long eclipse —
His name Urania writes with these,
And stamps it on her Pleiades.

An important influence in Benjamin Peirce's boyhood was that of Nathaniel Bowditch. The story of their early acquaintance is an amusing one. Bowditch's son Ingersoll showed his schoolmate Peirce a solution of a problem that the boys had been set to work out. After some consideration, young Benjamin pointed out an error real or conceived — and Ingersoll reported the matter to his father. "Show me the boy who corrects my mathematics," said Bowditch, and the acquaintance so begun quickly ripened into the relationship of master and follower. While in college, Peirce was in Bowditch's class and regularly read the proofs of his *"Mécanique céleste"*. Thirty years later, he dedicated his own masterpiece, the treatise on "Analytic Mechanics", "To the cherished and revered memory of my Master in Science, Nathaniel Bowditch, the father of American Geometry". The outward happenings of Peirce's life are soon told: born in Salem, Massachusetts, on April 4,

1809, he was graduated from Harvard in 1829 and then taught for two years in Northampton. In 1831 he returned to Harvard and remained there, first as tutor and later as professor, until his death in 1880, the longest association, save one, of any individual with that university. The story of his achievements, both as astronomer and mathematician, is not so quickly told. During his student days in Cambridge, Bowditch had predicted "that young Peirce would become one of the leading mathematicians of this century". At his death it was said of him that his merits ranked with the achievements of Bernouilli, Euler and Laplace. During the early days of his professorship he published a number of mathematical text-books covering the course then taught at the college, a "Treatise on Sound" on "Plane and Solid Geometry", "Treatise on Algebra", and a treatise on "Plane and Spherical Trigonometry", all of which greatly improved the then recognized methods of teaching mathematics. Their style was condensed and clear and substituted for the minute demonstrations of Euclid, the short, terse forms of geometry. Bowditch died in 1838, and his cloak fell upon the younger man, who stood chief among the frew who then had access to foreign journals. When the Harvard Observatory was founded a few years later on the occasion of the great comet of 1843, Peirce's share in the establishment was considerable. In 1841 he began a work on "Curves, Functions, and Forces" and in place of the third volume published his "Analytic Mechanics", which set forth the general principles and methods of the science as a branch of mathematical theory, and embodied in a systematic treatise the latest and best methods and forms of conceptions of the greatest geometers. This is perhaps the most characteristic of all his works, but its extreme brevity of expression makes it difficult reading for any but the initiate. He was requested to take charge of the mathematical part of the "American Almanac", and prepared ten volumes of it, in one of them publishing a list of the known orbits of comets to which he added several approximate orbits, computed by him, of great comets which had been only imperfectly observed. For it, too, he prepared his "Tables of the Moon" which were used for many years. After the discovery of Neptune in 1846, Peirce took the greatest interest in the researches of Leverrier and Adams, presenting to the Academy papers on such questions as: What is the orbit of the new planet? What is its mass? How much do they differ from the assigned orbits and masses? Perhaps his greatest contribution to astronomy were his discoveries regarding the rings of Saturn. He demonstrated that the rings, if solid, could not be sustained by the planet, that satellites could not sustain a solid ring, but that sufficiently large and numerous satellites could sustain a fluid ring, and that the actual satellites of Saturn were sufficient for that purpose. Professor Bache was superintendent of the Coast Survey, and in 1852 he induced Peirce to undertake the longitude determinations. In 1867 when Bache died, Peirce himself succeeded to the superintendency and for seven years did very active work for the organization, making tours of inspection, and raising the standard of service by giving great freedom of action to individual officers. He extended the scope of the Survey so as to carry geodetic work into interior states; as superintendent, took personal charge of the American expedition to Sicily to observe the eclipse of the sun in December, 1870; and in 1874 sent out parties to Nagasaki and to Chatham Island to take part in observations of the transit of Venus. Between

1864 and 1870 Peirce read before the National Academy of Sciences various papers, that were later published in his "Linear Associative Algebra"; which he called the "pleasantest mathematical effort of my life. In no others have I seemed to myself to have received so full a reward for my mental labor in the novelty and breadth of my results." During the last years of his life he published a series of eight propositions in cosmical physics, and delivered before the Lowell Institute a course of lectures on "Ideality in Science" which have been published in book form, and afford an interesting view of the speculative operations of his mind.

Bernoulli, Daniel. (b. Jan 29, 1700; Groningen, Netherlands; d. Mar. 17, 1782; Basel, Switzerland). Swiss mathematician. Daniel was a son of Johann I Bernoulli. Of all the Bernoulli family he was probably the most outstanding mathematician and certainly the one with the widest scientific interests. Daniel studied at the universities of Basel, Strasbourg, and Heidelberg. His studies, which reflected his already wide interests, included logic, philosophy, and medicine in addition to mathematics. In 1724 Daniel produced his first important piece of mathematical research — a work on differential equations, which sufficiently impressed the European scientific community to earn him an invitation to the St. Petersburg Academy of Sciences as a professor of mathematics. Once installed in Russia he continued to pursue his varied interests and obtained a post at the academy for his friend Leonhard Euler. In 1733 he left Russia to return to Switzerland to take up a chair in mathematics at Basel. Bernoulli's wide interests continued to occupy him and during his time at Basel he also held posts in botany, anatomy, physiology, and physics. In Switzerland Daniel did the work for which he is best known, namely his virtual founding of the modern science of hydrodynamics using Isaac Newton's laws of force. He published these ideas in his *Hydrodynamica* (1738). Apart from his work in fluid dynamics Daniel made distinguished contributions to probability theory and differential equations in mathematics, and to electrostatics in physics. He also laid the basis for the kinetic theory of gases. Like his uncle, Jakob I Bernoulli, Daniel corresponded voluminously with many scholars throughout Europe, thus extensively disseminating his new ideas.

Bernoulli, Jakob (or Jacques) I. (b. Dec. 27, 1654; Basel, Switzerland; d. Aug. 16, 1705; Basel). Swiss mathematician Jakob I was the first of the Bernoulli family of scientists to achieve fame as a mathematician. As with the two other particularly outstanding Bernoullis — his brother, Johann I, and nephew, Daniel — Jakob I played an important role in the development and popularization of the then recently invented integral and differential calculus of Isaac Newton and Gottfried Leibniz. His particular contribution to the calculus consisted in showing how it could be applied to a wide variety of fields of applied mathematics. Jakob I began studying theology and in 1676 traveled through Europe where he met many of the important scientists of the day, such as Robert Boyle in England. He returned to Basel in 1682 where he began lecturing on mechanics and held a chair in mathematics at Basel University from 1687 until his death. Apart from his mathematical work he was an influential figure in the European scientific community through his voluminous correspondence. His most important contribuitions to mathematics were in the fields of probability and in the calculus of variations. His work on probability is contained in his treatise the *Ars conjectandi* (1713; The Art of Conjecturing) in which he made numerous important contributions to the subject, among which was his discovery of what is now known as the law of large numbers. This work also contains Bernoulli's work on permutations and combinations. The Bernoulli family were always prone to rivalry and Jakob I and his younger brother, Johann I, became involved in a controversy over the problem of finding the shortest path between two points of a particle moving solely under the influence of gravity. The result of this vigorous dispute was the creation of the calculus of variations, a field that Leonhard Euler was later to develop. In addition to this Jakob I did important and useful work in the study of the catenary, which he applied to the design of bridges.

Bernoulli, Johann (or Jean) I. (b. Aug. 6, 1667; Basel, Switzerland; d. Jan. 1, 1748; Basel). Swiss mathematician Johann I was the brother of Jakob I Bernoulli (q.v.). As in the case of several of the Bernoulli family Johann I's father did not encourage him to make a career of mathematics and he graduated in medicine in 1694. Once he had abandoned medicine for mathematics he became chiefly interested in applying the calculus to physical problems. He played an important role as a propagandist for the calculus in general and in particular as a champion of Gottfried Leibniz's priority over Isaac Newton. Johann I held a chair in mathematics at Groningen, Holland, from 1695 and returned to Switzerland to take up a chair in mathematics at Basel on the death of his brother in 1705. Johann I's interests ranged over many fields outside mathematics including physics, chemistry, and astronomy. His mathematical work also included particularly important contributions to optics, to the theory of differential equations, and to the mathematics of ship sails.

Besicovitch, Abram Samoilovich. (b. Jan. 24, 1891; Berdjansk, now in the Soviet Union, d. Nov. 2, 1970; Cambridge, England). Soviet-British mathematician. Besicovitch graduated in 1912 from the University of St. Petersburg (now Leningrad), having studied under A.A. Markov. In 1917 he became professor of mathematics at Perm (later Molotov University) and then taught at the Leningrad Pedagogical Institute and Leningrad University. He left the Soviet Union in 1924 to work briefly with Harold Bohr in Copenhagen before moving to England. G.H. Hardy was sufficiently impressed by Besicovitch's analytical powers to secure him a lectureship at Liverpool University. After a year he became a lecturer at Cambridge and in 1950 became Rouse Ball Professor of Mathematics, a post that he held for eight years. Besicovitch was primarily an analyst. The subject with which he is most associated is that of almost periodic functions, an interest stemming from his collaboration with Bohr. 1932 saw the publication of his book *Almost Periodic Functions.* He also did important work on the 'Karkeya problem', general real analysis, complex analysis, and various geometric problems. His other main work was on geometric measure theory.

Blaise Pascal. A seven-year old boy was busy making geometrical figures on the ground. He was so absorbed in his play that he was not aware of the surroundings. Suddenly, his father came in and scolded him. He snatched his books of mathematics. The father thought a boy of his age should not study a drab and difficult subjects like mathematics. But the boy continued his mathematical studies, and made some conclusions. At the age of 12, he proved before his father that the sum of the three angles of a triangle is equal to two right angles. His father was impressed and allowed the boy to study mathematics freely. The name of this boy was Blaise Pascal, who, at the age of 12, proved this theorem which has become a universal theorem of geometry today. At the age of 17, he published an essay relating to mathematics which was admired by scientist like Descartes. Pascal was a well-known mathematician, physicist, philosopher and religious writer. Perhaps, there may not be any student who has not studied Pascal's law. According to this law when pressure is applied at any point in a fluid, it gets transmitted equally in all directions. On the basis of this law, the inventions of syringe, hydraulic press and hydraulic brakes became possible. Pascal was born and brought up in France. He invented a triangle called Pascal's triangle which consists of rows of numbers arranged in a certain way. This triangle is shown in the figure below. It has proved very useful in the study of probability. Pascal's father was an accountant in the local administration. He used to do official work at home even late at night. Pascal was very unhappy about it. To help his father, he decided to develop a calculating machine which would reduce the pressure on his father in mathematical calculation. He invented a calculating machine which was operated by gear wheel. This machine could do the operations of addition and subtraction. It helped his father a great deal. He got this machine patented, but being costly it could not be manufactured on large scale. Based on Pascal's model, the first commercial calculating machine was made by American engineer William Burrughs in 1892. Pascal did a lot of work on geometry, probability, hydrostatics, integral calculus etc. He wrote several religious books also. Early in 1659, Pascal fell seriously ill, and in 1662 he died.

Birkhoff, George David. (1884-1944), American analyst and topologist who was President of both the American Mathematical Society and the Association for the Advancement of Science, and influenced an entire generation of American mathematicians. Although his own major work was in the application of analysis to dynamics, he also contributed to the study of difference equations, constructed a relativistic theory of gravity independently of Einstein, and devised a mathematical theory of 'aesthetic measure'. Birkhoff studied at the Lewis Institute (now the Illinois Institute of Technology) from 1896 to 1902, and subsequently at the University of Chicago and at Harvard. In 1907 he obtained his PhD from Chicago and took up a teaching post at the University of Wisconsin, moving to Princeton in 1909. In 1912 he became assistant professor at Harvard and, in 1919, professor there, a post he held until 1939. Birkhoff's mathematical interests were wide, and among the many areas to which he made notable contributions were differential equations, celestial mechanics, difference equations, and the three-body problem. His main field of research was mathematical analysis, especially applied to dynamics. In the course of his work on dynamical systems Birkhoff obtained a famous proof of a conjecture made by Henri Poincare in topology, usually known as Poincare's last geometric theorem. The ergodic theorem, a result concerned with the formal mathematics of probability theory, that Birkhoff proved in 1931, is another of his outstanding achievements. Modern dynamics received an enormous impetus from Birkhoff's work, and he also worked on the foundations of relativity and quantum mechanics.

Bogoliubov, Nikolai Nikolaevich. (b. Aug. 5, 1909; Nizhny Novgorod, now in the Soviet Union). Soviet mathematician and physicist. Bogoliubov was accepted for graduate work at the Academy of Sciences of the Ukranian SSR, and subsequently worked there and at the Soviet Academy of Sciences. His main contribution has been in the application of mathematical techniques to theoretical physics. He developed a method of distribution-function for non-equilibrium processes, and has also worked on superfluidity, quantum field theory, and superconductivity. His work was partly paralleled by the work of John Bardeen, Leon Cooper, and John Robert Schrieffer. Bogoliubov has been a prolific author, with many of his works translated into English. He has also been active in founding scientific schools in nonlinear mechanics, statistical physics, and quantum-field theory. Since 1963 he has been academician-secretary of the Soviet Academy of Sciences and was made director of the Joint Institute for Nuclear Research in Dubna in 1965. The next year he was also made a deputy to the Supreme Soviet. He has been honored by scientific societies throughout the world, and in his own country has received the State Prize twice (1947 and 1953) and the Lenin Prize (1958).

Boltzmann, Ludwig Edward. (b. Feb. 20, 1844; Vienna; d. Sept. 5, 1906; Duino, now in Italy). Austrian theoretical physicist. Boltzmann studied at the University of Vienna, where he received his doctorate in 1866. He held professorships in physics or mathematics at Graz (1869-

73; 1876-79), Vienna (1873-76; 1894-1900; 1902-06), Munich (1889-93), and Leipzig (1900-02). Boltzmann made important contributions to the kinetic theory of gases. He developed the law of equipartition of energy, which states that the total energy of an atom or molecule is, on average, equally distributed over the motions (degrees of freedom). He also produced an equation showing how the energy of a gas was distributed among the molecules (called the *Maxwell-Boltzmann distribution)*. Boltzmann also worked on thermodynamics, in which he developed the idea that heat, entropy, and other thermodynamic properties were the result of the behavior of large numbers of atoms, and could be treated by mechanics and statistics. In particular, Boltzmann showed that entropy — introduced by Rudolf Clausius — was a measure of the disorder of a system. *Boltzmann's equation* (1896) relates entropy (S) to probability (p): S = klog/p + b. The constant k is known as *Boltzmann's constant* and has the value 1.38054×10^{-23} joule per kelvin. The equation is engraved on his gravestone. Boltzmann's work in this field was heavily criticized by opponents of atomism, particularly Wilhelm Ostwald. It did however lead to the science of statistical mechanics developed later by Josiah Willard Gibbs and others. Boltzmann also worked on electromagnetism. He is further noted for a theoretical derivation of the law of radiation discovered by Josef Stefan (q.v.). Toward the end of his life he suffered from illness and depression, and committed suicide in 1906.

Bolyai, Janos. (b. Dec. 15, 1802; Koloszvár, now Cluj in Rumania; d. Jan. 27, 1860; Marosvásárhely, now Tirgu-Mures in Rumania). Hungarian mathematician. Bolyai's father Farkas was a distinguished mathematician who had an obsession with the status of Euclid's famous parallel postulate and devoted his life to trying to prove it. Despite his father's warnings that it would ruin his health, peace of mind, and happiness, Janos too started working on this axiom until, in about 1820, he came to the conclusion that it could not be proved. He went on to develop a consistent geometry in which the parallel postulate is not used, thus establishing the independence of this axiom from the others. Bolyai published an account of his non-Euclidean geometry in 1832. Although his discovery had been anticipated by Nikolai Lobachevsky (q.v.) and Karl Gauss he was unaware of their work. The discovery of the possibility of non-Euclidean geometries had a tremendous impact on both mathematics and philosophy. In mathematics it opened the way for a far more general and abstract approach to geometry than had previously been pursued, and in philosophy it settled once and for all the arguments about the supposed privileged status of Euclid's geometry. Bolyai also did valuable work in the theory of complex numbers.

Bondi, Sir Hermann. (b. Nov. 1, 1919; Vienna). British-Austrian mathematician and cosmologist. Bondi studied at Cambridge University, where he later taught. In 1954 he moved to London to take up the chair in mathematics at King's College. Bondi has always been actively interested in the wider implications of science and the scientific outlook, as his membership of the British Humanist Association and the Science Policy Foundation testify. He has served as chief scientific adviser to the Ministry of Defence (1971-77), chief scientist at the Department of Energy (from 1977), and chairman of the Natural Environment Research Council (from 1980). He was knighted in 1973. Bondi's most important work has been in applied mathematics and especially in cosmology. In collaboration with Thomas Gold, he propounded, in 1948, a new version of the steady-state theory of the universe. This, among other topics, forms the substance of Bondi's book *Cosmology* (1952). The idea of a steady-state theory had first been suggested by Fred Hoyle, the purpose being to devise a model of the universe that could accommodate both the fact that the universe is the same throughout, and yet is expanding. Bondi and Gold's model was innovatory in postulating that there is continuous creation of matter in order to maintain the universe's homogeneity despite its expansion. Although it enjoyed considerable popularity, Bondi and Gold's steady-state model is now considered to have been decisively refuted by observational evidence and the big-bang theory is favored.

Boole, George. (b. Lincoln, UK, 2.11.1815; d. Ballintemple, Munster, Ireland, UK [now Irish Republic], 8.12.1864). British mathematician and logician. Boole learned his early lessons in mathematics from his father, an amateur mathematician and optical instrument maker. Thereafter he was self-taught. He became a school teacher at 16. In 1847 he published a pamphlet *The Mathematical Analysis of Logic* (Cambridge) which bridged the gap previously separating mathematics from formal logic. This development was crucial in advancing the potential powers of the analytical engines of Babbage, and later of the 20c digital computers. Boole became a protége of the English mathematician Augustus de Morgan, and with his support was elected to the chair of mathematics at Queen's College, Cork, in spite of his lack of formal qualifications. In 1854 he published his mature work on mathematical logic *An Investigation into the Laws of*

Thought, on which are Founded the Mathematical Theories of Logic and Probabilities (London). In 1859 he published his *Treatise on Differential Equations* (Cambridge) and in 1860 his *Treatise on the Calculus of Finite Differences* (Cambridge). The development of Boolean algebra was fundamental to mathematical logic, and is the basic logical tool in designing modern computers. William Stanley Jevons built a 'logical piano' for implementing Boolean algebra in the 1860s. His *Collected Logical Works* (London) were published in 1916. Boole came from a poor background and was virtually self-taught in mathematics. He discovered for himself the theory of invariants. Before he obtained an academic post Boole spent several years as a school teacher, first in Yorkshire and later at a school he opened himself. In 1849 he became professor of mathematics at Queen's College, Cork, Ireland. Boole's main work was in showing how mathematical techniques could be applied to the study of logic. His book *The Laws of Thought* (1854) is a landmark in the study of logic. Boole laid the foundations for an axiomatic treatment of logic that proved essential for the further fundamental developments soon to be made in the subject by such workers at Gottlob Frege and Bertrand Russell. Boole's own logical algebra is essentially an algebra of classes, being based on such concepts as complement and union of classes. His work was an important advance in considering algebraic operations abstractly — that is, studying the formal properties of operations and their combinations without reference to their interpretation or 'meaning'. Fundamental formal properties like commutativity and associativity were first studied in purely abstract terms by Boole. Boole's work led to the recognition of a new and fundamental algebraic structure the *Boolean algebra* alongside such structures as the field, ring, and group. The study of Boolean algebras both in themselves and their application to other areas of mathematics has been an important concern of 20th-century mathematics. Boolean algebras find important applications in such diverse fields as topology, measure theory, probability and statistics, and computing.

Borel, Félix Edouard Justin Emile. (1871-1956), French measure theorist and probability theorist, who together with Lebesgue and Baire founded the theory of real-valued functions, and also contributed to the development of game theory. He was also a member of the Chamber of Deputies and for fifteen years was Navy Minister until his imprisonment by the Vichy regime; he then joined the French Resistance. He was appointed to a Chair specially created for him at the Sorbonne in 1909, received the *Croix de Guerre* after the First World War, and the Resistance Medal and Grand Cross of the *Légion d'Honneur* after the Second, and was the first gold medalist of the French National Centre for Scientific Research in 1955. His mathematical activity at first centred on problems in mathematical analysis, especially the growing use then being made of the set theory of Cantor. One of Borel's theorems of that time is now normally called 'the Heine-Borel theorem'. He also systematized the study of nonconvergent infinite series, and obtained many results for functions of a complex variable. An important aspect of these achievements was to individuate the concept of a 'measure' as a generalization of the intuitive ideas of length, breadth and volume. In the 1900s he noticed that the properties of a measure were also satisfied by probability, and worked extensively on that subject from then on. He was also a founder of game theory. Borel took an interest in the philosophy of mathematics. He put forward a form of 'constructivism'. which asserts that only processes which actually indicate the method of construction are valid in mathematics. In addition to his research and administrative duties, Borel was extremely active as a popularizer and textbook writer in mathematics. He edited, and contributed to, an important series of mathematical monographs for the house of Gauthier-Villars. His writings are collected in *Oeuvres* (4 vols. Paris, 1972). With his wife, the novelist Camille Marbo, he launched in 1906 the general periodical *La Revue du mois,* which ran for 14 years. His public positions included minister of the navy (1925) and founder-director of the Institute Henri Poincaré from 1928 until his death.

Borelli, Glovanni Alfonso. (b. Jan. 28, 1608; Naples, Italy; d. Dec. 31, 1679; Rome). Italian mathematician and physiologist. Borelli's mathematical training — he was professor at Messina and Pisa — led him to apply mathematical and mechanical laws to his two main interests, astronomy and animal physiology. He rightly explained muscular action and the movements of bones in terms of levers, and also carried out detailed studies of the flight mechanism of birds. However, his extension of such principles of internal organs, such as the heart, stomach, and lungs, overlooked the essential chemical actions that take place in these organs. Borelli's *De motu animalium* (1680; On the Movement of Animals), which includes his theory of blood circulation, is thus in part erroneous. In astronomy, he tried to explain the motion of Jupiter's planets by postulating that Jupiter, as well as the Sun, exerts a gravitational pull. He also emphasized the findings of Jhoannes Kepler and Jeremiah Horrocks in drawing attention to the elliptical orbits of the Sun, Moon, and the planets and was the first to suggest that comets trace a parabolic path through the solar system.

Bouguer, Pierre. (b. Feb. 16, 1698; Le Croisic, France; d. Aug. 15, 1758; Paris). French physicist and mathematician. Bouguer's father was a hydrographer and mathematician and Bouguer followed him into the same profession. He was a child prodigy and obtained a post as professor of hydrography at the remarkably early age of 15. The study of the problems associated with navigation and ship design was his chief interest. Bouguer took part in an extended expedition to Peru led by Charles de la Condamine to determine the length of a degree of the meridian near the equator. While on this expedition Bouguer also did a great deal of other valuable experimental work. One of Bouguer's most successful inventions was the heliometer to measure the light of the Sun and other luminous bodies. Although it was not his chief interest the research for which Bouguer is now best remembered was on photometry. Here too he did much valuable experimental work and one of his major discoveries was of the law now named for him. This states that in a medium of uniform transparency the intensity of light remaining in a collimated beam decreases exponentially with the length of its path in the medium. The law is sometimes unjustly attributed to Johann Lambert. Bouguer's work in optics can be seen as the beginning of the science of atmospheric optics.

Bourbaki, Nicolas. Collective *nom de plume* under which a group of (mostly) French mathematicians publishes an encyclopedic treatise on advanced mathematics called *Elements de mathématiques.* It was launched in the early 1930s by mathematicians such as H. Cartan, J.

Dieudonne and Andre Weil, who had found themselves, as students in the 1920s, taught by much older men, the intermediate generation having been killed during WWI. They wanted to create a *published* record of mathematics, which would be immune from the perils of mass carnage in the future. Retirement from the group at 50 is compulsory, in order to prevent Bourbakist mathematics from becoming *passé*. Nevertheless, the modern generation of young French mathematicians is mostly unsympathetic to Bourbaki's rather arid and formal style, in which great emphasis is laid on set theory and abstract algebra, and theories and proofs are formalized to a high degree of sophistication. The *Elements* began appearing in earnest after WW2, but have become less frequent in recent years. 'Bourbaki' is the collective *nom de plume* of a group of some of the most outstanding of contemporary mathematicians. The precise membership of Bourbaki, which naturally has changed over the years, is a closely guarded secret but it is known that most of the members are French. Since 1939, Bourbaki has been publishing a monumental work, the *Elements de mathematique* (Elements of Mathematics), of which over thirty volumes have so far appeared. In this Bourbaki attempts to expound and display the architecture of the whole mathematical edifice starting from certain carefully chosen logical and set-theoretic concepts. The emphasis throughout the *Elements* is on the interrelationships to be found between the various structures present in mathematics, and to a certain extent this means that Bourbaki's exposition cuts across traditional boundaries, such as that between algebra and topology. Indeed for Bourbaki, pure mathematics is to be thought of as nothing other than the study of pure structure. Since the members of Bourbaki are all working mathematicians, rather than pure logicians, in contrast to other foundational enterprises (e.g. those of Gottlob Frege, Bertrand LRussell, and A.N. Whitehead) the influence of Bourbaki's writings on contemporary mathematicians and their conception of the subject has been immense.

Box, George Edward Pelham. (b. Gravesend, Kent, UK, 18.10.1919), British statistician. Employed at ICI from 1948 to 1956, he has since then pursued an academic career in the USA. In statistics, modelling of mechanisms in the light of data is commonly a process that uses, and potetially rejects, all existing theory. This is tiresome for outside users and impossible for computers. Box has a reputation for finding areas in which, and procedures by which, this modelling can be made highly systematic: especially (with G.M. Jenkins) in the forecasting of time-series, where his techniques (partly for lack of alternatives) have been very widely adopted (see *Time-series Analysis, Forecasting, and Control,* NY, 1970). He also produced (with G. Tiao) an analytical system for scientific experiemnts using subjective probabilities.

Brahmagupta. (b. c. 598; d. c. 665) Indian mathematician and astronomer. Brahmagupta was the director of the observatory at Ujjain, a town in central India. In 628 he wrote the *Brahmasphuta-siddhanta* (The Opening of the Universe) half of which is about mathematics and half about astronomy. He introduced negative numbers into India, gave a satisfactory rule for the solution of quadratic equations, and attempted to apply algebra to astronomy. He discovered the formula for the area of a cyclic quadrilateral: A = /[(s - a)(s - b)(s - c)(s - d)] where s is half the perimeter and a, b, c, and d are the lengths of the sides. One innovation introduced by him was the use of oval epicycles for Mars and Venus. He rejected the view that the stars are stationary and the Earth moves, on the grounds that any such movement would cause high buildings to fall.

Brans, Carl Henry. (b. Dec. 13, 1935; Dallas, Texas) American mathematical physicist. Brans graduated in 1957 from Loyola University, Louisiana, and obtained his PhD in 1961 from Princeton. He returned to Loyola in 1960 and in 1970 was appointed professor of physics. Brans has worked mainly in the field of general relativity. He is best known for his production with Robert Dicke (q.v.) in 1961 of a variant of Einstein's theory in which the gravitational constant varies with time. A number of very accurate measurements made in the late 1970s has failed to detect this and some of the other predictions made by the *Brans-Dicke theory*.

Bridgman, Percy Williams. (b. Apr. 21, 1882; Cambridge, Massachusetts; d. Aug. 20, 1961; Randolph, New Hampshire) American mathematician and physicist. Bridgman, the son of a journalist, was educated at Harvard where he obtained his PhD in 1908. He immediately joined the faculty, leaving only on his retirement in 1954 after serving as professor of physics from 1919 to 1926, professor of mathematics and natural philosophy from 1926 to 1950, and as Huggins Professor from 1950 to 1954. Most of Bridman's research has been in the field of high-pressure physics. When he began he found it necessary to design and build virtually all his own equipment and instruments. In 1909 he introduced the self-tightening joint and with the appearance of high tensile steels he could aim for pressures well beyond the scope of earlier workers. At the beginning of the century Emile Amagat and Louis Cailletet had attained pressures of some 3000 kilograms per square centimeter; Bridgman increased this enormously, regularly attaining pressures of 100 000 kg/cm². Bridgman used such pressures to explore the properties of numerous liquids and solids. In the course of this work he discovered two new forms of ice, freezing at temperatures above 0°C. He also, in 1955, transformed graphite into synthetic diamond. Bridgman was awarded the 1946 Nobel Physics Prize for his work on extremely high pressures. He was also widely known as a philosopher of science and in his book *The Logic of Modern Physics* (1927) formulated his theory of 'operationalism' in which he argued that a concept is simply a set of operations. In his 70s Bridgman developed Paget's disease, which gave him considerable pain and little prospect of relief. He committed suicide in 1961.

Briggs, Henry. (b. February 1561; Warley Wood, England; d. Jan. 26, 1630; Oxford, England). English mathematician. Briggs became a fellow of Cambridge University in 1588 and was later made a lecturer (1592) and a professor (1596) of geometry at Gresham College, London. He is remembered chiefly for the modifications he made to John Napier's (q.v.) logarithms, which were first published in 1614. Napier had produced these to base e (natural logarithms) but Briggs considerably improved their convenience of use by introducing the base 10 (common logarithms). He also introduced the modern method of long division. Briggs became Savilian Professor of Geometry at Oxford in 1619.

Brouncker, William Viscount. (b. 1620; d. Apr. 5, 1685; London). English mathematician and experimental scientist. Brouncker graduated from Oxford University in 1647 with a degree in medicine. He held a variety of official posts, including serving as member of parliament and president of Gresham's College. He was a friend of the eminent mathematician John Wallis and his own most notable work was also in mathematics. He was a founder and first president of the Royal Society (1662-77) and as such carried out experimental work. Brouncker usually contented himself with solving problems arising from the work of other mathematicians rather than doing creative work himself but was the first to use continued fractions. He was a friend of Samuel Pepys and frequently figures in Pepys's *Diary*. Apart from science Brouncker had a lively interest in music.

Brouwer, Luitzen Edbertus Jan. (b. Overschie, S. Holland, The Netherlands, 27.7.1881; d. Blaricum, N. Holland, 2.12.1966). Dutch mathematician. Brouwer's chief mathematical interests were twofold. One was a philosophical standpoint about mathematics, in which he permitted only methods of reasoning which led by direct construction from step to step. He called this view 'intuitionism', since he considered that 'intuitionist mathematics is an essentially languageless activity of the mind having its origin in the perception of a move of time'. Brouwer's position was controversial for several reasons. Firstly, it was very radical; for example, it entailed the rejection of the law of excluded middle. Secondly, his assertion of the primacy of time in mathematical thought was philosophically contentious (one could say, naive). Thirdly, he brought into mathematics elements of mysticism (on which he published a book) by allowing a role for languagelessness (akin to states of mystical ecstasy). Finally, his technical reconstruction of mathematics was full of difficult definitions and obscure procedures. Nevertheless, his criticism of established mathematical procedures deepened mathematicians' awareness of philosophical problems in their subject and has led to the development of similar, and less cryptically phrased, positions. Brouwer's second interest lay in topology, where he not only proved various important theorems but also exposed the naivety of the ordinary conception of dimension. His work here involved important applications of the set theory of CANTOR. His writings are published as *Collected Works* (2 vols, Amsterdam, 1975-6)

Brouwer, Luitzen Egbertus Jan. (b. Feb. 27, 1881; Overschie, Netherlands; d. Dec. 2, 1966; Blaricum, Netherlands). Dutch mathematician and philosopher of mathematics. Brouwer took his first degree and doctorate at the University of Amsterdam, where he became successively *Privatadozent* and professor in the mathematics department. From 1903 to 1909 he did important work in topology, presenting several fundamental results, including the fixed-point theorem. This is the principle that, given a circle (or sphere) and the points inside it, then any transformation of all points to other points in the circle (or sphere) must leave at least one point unchanged. A physical example is stirring a cup of coffee—there will always be at least one particle of liquid that returns to its original position no matter how well the coffee is stirred. Brouwer's best-known achievement was the creation of the philosophy of mathematics known as *intuitionism*. The central ideas of intuitionism are a rejection of the concept of the completed infinite (and hence of the transfinite set theory of Georg Cantor) and an insistence that acceptable mathematical proofs be constructive. That is, they must not merely show that a certain mathematical entity (e.g. a number or a function) exists, but must actually be able to construct it. This view leads to the rejection of large amounts of widely accepted classical mathematics and one of the three fundamental laws of logic, the law of excluded middle (either p or not-p; a proposition is either true or not true). Brouwer was able to re-prove many classical results in an intuitionistically acceptable way, including his own fixed-point theorem.

Bullen, Keith Edward. (b. June 29, 1906; Auckland, New Zealand; d. Sept. 23, 1976; Sydney, Australia). Australian applied mathematician and geophysicist. Bullen, the son of Anglo-Irish parents, was educated at the universities of Auckland, Melbourne, and Cambridge, England. He began his career as a teacher in Auckland then lectured in mathematics at Melbourne and Hull, England. In 1946 he became professor of applied mathematics at the University of Sydney. Bullen made his chief contributions to science from his mathematical studies of earthquake waves and the ellipticity of the Earth. In 1936 he gave values of the density inside the Earth down to a depth of 3100 miles (5000 km). He also determined values for the pressure, gravitation intensity, compressibility, and rigidity throughout the interior of the Earth as a result of his mathematical studies. From the results on the Earth's density he inferred that the core was solid and he also applied the results to the internal structure of the planets Mars, Venus, and Mercury and to the origin of the Moon. Bullen conducted some of his early work in collaboration with Harold Jeffreys on earthquake travel times. This resulted in the publication of the *Jeffreys-Bullen* (JB) *tables* in 1940.

C

Cantor, Georg Ferdinand Ludwig Philipp. (b. St. Petersburg [now Leningrad], Russia, 3.3.1845, naturalized German citizen; d. Halle, Germany, 6.1.1918). Russian/German mathematician. Educated at Zürich, Berlin and Göttingen universities and appointed to the University of Halle in 1869. Cantor was responsible for the development of modern set theory, the system to which all modern mathematics is referred. His first achievements were in the traditional field of trigonometric series, whence he was led to consideration of the different kinds of sets of numbers that can appear in mathematics. He gradually showed that 19c ideas on the connection between the dimension of a set and the number of its elements were dangerously fallacious. For instance, the plane, the real continuum, and the Cantor ternary set all have the same number of elements, although the first covers two dimensions; the second, one; and the third, nothing. All of them are uncountable using the natural numbers, and Cantor, while inventing many concepts in point-set topology (closure, denseness) which have been essential to successful theories of dimension, concentrated on enumeration. He used the concept of the one-to-one mapping as his equivalence relation, and developed the ordinal numbers as the equivalence classes when these mappings are order-preserving. Thus if the natural numbers (1, 2, 3, ...) are the first ordinal, the set (2, 3, 4,

...; 1) is the second, since the numbers 2, 3, ... 'absorb' all the numbers in the first set Cardinal numbers are obtained from ordinal numbers by relaxing the restriction; both the sets cited have the same cardinality. Cantor showed that the set of all sequences of elements from a set of given cardinality has a higher cardinality than the parent set, but could not prove his hypothesis that there is no cardinal number between that of the natural numbers and the cardinality of the continuum. This continuum hypothesis has been proved, after great labour, to be optional. Cantor assumed that all sets could be well-ordered, that is, arranged so that every subset has a least element within itself; this postulate, which is equivalent to the axiom of choice, provoked great dissension among mathematicians then as now. It is logically independent of the other axioms needed for formalizations of set theory; most pure mathematicians find it useful; but no one has the least conception of what a well-ordering of the real continuum would look like.

Cantor, Georg. (b. Mar. 3, 1845; St. Petersburg, now Leningrad in Russia; d. Jan. 6, 1918; Halle, now in East Germany). German mathematician. Cantor was born of (Christian) Danish parents of Jewish descent, who moved to Germany in 1856. He studied at the universities of Zurich, Berlin, and Göttingen and taught at the University of Halle. In addition to mathematics he was deeply interested in philosophy and theology. Rather surprisingly he also took astrology seriously. Cantor began his mathematical work in the field of analysis, but his interest in the sets of points of convergence of Fourier series led him to the work for which he is remembered — the creation of the theory of transfinite sets. For centuries the concept of infinity had been a highly controversial one both mathematically and philosophically. Before Cantor a mathematician of the stature of Karl Friedrich Gauss could declare that all talk of the 'completed infinite' was a dangerous confusion that had no place in mathematics. What Cantor did was to show that the concepts of cardinal and ordinal number could be defined mathematically in such a way that it made perfectly clear sense to talk of infinite or transfinite numbers. Cantor's mathematics of the transfinite met initially with a very hostile response from mathematicians. A prominent exception was Richard Dedekind, who recognized the value of Cantor's work from the start. The mathematician Leopold Kronecker launched a particularly vitriolic attack on Cantor that may well have contributed to the decline in Cantor's mental health. He actually died in a mental hospital. Cantor did not present his set theory in the completely rigorous way that is now customary since he took the notion of 'set' to be intuitively clear. But the discovery of set-theoretic paradoxes by Cantor himself, bu Bertrand Russell, and others showed that some refinement was needed. This led to the development of axiomatic set theory by Ernst Zermelo and others.

Caratheodory, Constantin. (1873-1950), German analyst who worked as an engineer in Egypt before studying mathematics, and later taught in Germany, Poland, and Greece. He saved the library of the new Greek University of Smyrna from the Turks and transferred it to Athens. His most important work was in the CALCULUS OF VARIATIONS, but he also made significant contributions to the theory of functions of several variables, measure theory, thermodynamics, and relativity.

Cardano, Gerolamo. (b. Sept. 24, 1501; Pavia, Italy; d. Sept.

21, 1576;Rome). Italian mathematician, physician, and astrologer. The work of Cardano constitutes a landmark in the development of algebra and yet in his own time he was chiefly known as a physician. He studied medicine at the universities of Pavia and Padua, receiving his degree in 1526, and spent much of his life as a practicing physician. He became professor of medicine at Pavia in 1543 and one of his notable nonmathematical achievements was to give the first clinical description of typhus fever. It was however in mathematics that Cardano's real talents lay. His chief work was the *Ars magna* (1545; The Great Skill) in which he gave ways of solving both the general cubic and the general quartic. This was the first important printed treatise on algebra. The solution of the general cubic equation was revealed to him by Nicolo Tartaglia in confidence and Cardano's publication aroused a bitter controversy between the two. Cardano's former servant, Lodovico Ferrari, had discovered the solution of the general quartic equation. In his later *Liber de ludo aleae* (Book on Games of Chance) Cardano did some pioneering work in the mathematical theory of probability. Cardano's interests were not, however, limited to mathematics and medicine. He also indulged in philosophical and astrological speculation and this had the unfortunate consequence that in 1570 he was charged with heresy by the Church. He was briefly jailed but was soon released after the necessary recantation. As a result of this episode Cardano lost his post as a professor at the University of Bologna, which he had held since 1562.

Cartan, Elie. (b. Dolomieu, Isère, France, 9.4.1869; d. Paris, 6.5.1951). French mathematician. Cartan was decisive in extending the theory of analysis on differentiable manifolds, now perhaps the central domain of mathematics. Cartan's thesis (1894) greatly clarified and sometimes corrected work of Killing and M. Lie, classifying simple, real and complex Lie algebras. Later he studied their representations and in 1913 discovered the spinor groups, of use in quantum mechanics. After 1925, influenced by WEYL, he studied the global theory of compact Lie groups and outlined their homology theory. He studied systems of differential equations in great generality, and he prefigured the theory of infinite Lie pseudo-groups. His work on differential geometry was decisive in its development to conceptual generality; in particular the concept of fibre bundle is implicit in it, and his theory of connections (analogues of directional derivatives on manifolds) which he discussed in correspondence with EINSTEIN is fundamental in general relativity as presently formulated. Cartan was professor of mathematics at Paris University from 1912 to 1940. Until 1930, however, he was somewhat isolated by virtue of the profound novelty of his interests, and also because of the weak state of French mathematics following WWI in which several young French specialists had been killed. His work was edited as *Oeuvres complètes* (6 vols, Paris, 1952-5).

Cartan, Elie Joseph. (b. Apr. 9, 1869; Dolomieu, France; d. May 6, 1951; Paris). French mathematician. Cartan is now recognized as one of the most powerful and original mathematicians of the 20th century, but his work only became widely known toward the end of his life. Cartan studied at the Ecole Normale Supérieure in Paris, and held teaching posts at the universities of Montpellier, Lyons, Nancy, and from 1912 to 1940, Paris. Cartan's most significant work was in developing the concept of analysis on differentiable manifolds, which now occupies a central place in mathematics. He began his research career with a dissertation on Lie groups — a topic that led him on to his pioneering work on differential systems. The most import innovation in his work on Lie groups was his creation of methods for studying their global properties. Similarly his work on differential systems was distinguished by its global approach. One of his most useful inventions was the 'calculus of exterior differential forms', which he applied to problems in many fields including differential geometry, Lie groups, analytical dynamics, and general relativity. Cartan's son Henri is also an eminent mathematician.

Cauchy, Baron Augustin Louis. (b. Aug. 21, 1789; Paris; d. May 23, 1857; Sceaux, France). French mathematician. Cauchy showed great mathematical talent at an early age and came to the attention of Joseph Lagrange and Pierre Laplace, who encouraged him in his studies. He was educated at the Ecole Polytechnique, where he later lectured and became professor of mechanics in 1816, and worked briefly as an engineer in Napoleon's army. He held extreme conservative views in religion and politics, typical of which was the strong allegiance to the Bourbon dynasty that caused him to follow Charles X (who had ennobled Cauchy) into exile in 1830. Cauchy then became professor of mathematics at Turin, but returned to France in 1838 and resumed his post at the Ecole Polytechnique. Cauchy was an extremely prolific mathematician who made oustanding contributions to many branches of the subject, ranging from pure algebra and analysis to mathematical physics and astronomy. He was also an outstanding teacher. His greatest achievements were in the fields of real and complex analysis in which he was one of the first mathematicians to insist on the high standards of rigor now taken for granted in mathematics. He gave the first fully satisfactory definitions of the fundamentally important concepts of limit and convergence. In particular, he laid the foundations of modern analysis in terms of limits and continuity, and developed the theory of functions of complex variables. After serving as an engineer in the force preparing Napoleon's abortive invasion of Britain, he was encouraged to pursue a mathematical career by Laplace (whom he had met when his family fled from the Reign of Terror) and by Lagrange. He became Professor at the *École Polythenique,* the Sorbonne, and the *Collège de France.* Because of his political and religious views, he refused to take the oath of allegiance to Louis-Philippe in 1830 and followed Charles X into exile; the University of Turin appointed him to a specially created Chair, but he left to tutor the grandson of Charles X. He published a total of 789 works, including monographs on definite integrals and on wave propagation, and papers on geometry, number theory, elasticity, the theory of error, astronomy, and optics.

Cavalieri, Francesco Bonaventura. (b. 1598; Milan, Italy; d. Nov. 30, 1647; Bologna, Italy) Italian mathematician and geometer. Cavalieri joined the Jesuits as a boy. He became interested in mathematics while studying Euclid's works and met Galileo, whose follower he became. Cavalieri's fame rests chiefly on his work in geometry in which he paved the way for the development of the integral calculus by Isaac Newton and Gottfried Leibniz. In 1629 Cavalieri became professor of mathematics at Bologna, a post he held for the rest of his life. At Bologna he developed his 'method of indivisibles', published in his *Geometria indivisibilibus continuorum nova quadam ratione promota* (1635: A Certain Method for the Development of a New Geometry of Continuous Indivisibles), which had much in common with the basic ideas of integral calculus. Cavalieri also helped to popularize the use of logarithms in Italy through the publication of his *Directorium generale uranometricum* (1632: A General Directory of Uranometry).

Cayley, Arthur. (b. Aug. 16, 1821; Richmond, England; d. Jan. 26, 1895; Cambridge, England). British mathematician. Cayley studied mathematics at Cambridge University, but before becoming a professional mathematician he spent 14 years working as a barrister. He was forced to do this since he was unwilling to take holy orders — at that time a necessary condition of continuing his mathematical career at Cambridge. When this requirement was dropped, Cayley was able to return to Cambridge and in 1863 became Sadlerian Professor there. Cayley was an extremely prolific mathematician. His greatest work was the creation of the theory of invariants, in which he worked closely with his friend James Joseph Sylvester. Cayley developed this theory as a branch of pure mathematics but it turned out to play a crucial role in the theory of relativity, as it is important in the calculation of space-time relationships in physics. He also developed the theory of matrices and made major contributions to the study of n-dimensional geometry. He went a considerable way toward unifying the study of geometry. Cayley also did important work in the theory of elliptic functions. One of Cayley's notable nonmathematical achievements was playing a large role in persuading the University of Cambridge to admit women as students.

Chang Heng. (b. AD 78; d. 142). Chinese astronomer, mathematician, and instrument maker. Chang Hengh was the Astronomer Royal at the court of the emperors of the Later Han. Although none of his works have survived there are detailed reports of his achievements, which are, by any standard, numerous and impressive. As a mathematician he is reported to have given 3.1622 or the square root of 10 as the value of p, which was as good as any other attempt of that period, apart from that of Archimedes. In astronomy he gave a detailed description of the figure of the universe in which the Earth lies at the centre of a large sphere like the yolk of an egg, which was an improvement on the earlier conception of a hemispherical heaven standing over the Earth like an umbrella. His real originality however lay in the introduction and design of scientific instruments. Thus he introduced a complete armillary sphere at about the same time as his western contemporary, Ptolemy. This was used to determine positions of celestial bodies and consisted of an interconnected set of such main circles of the celestial sphere as the equator, ecliptic, horizon, and meridian. Chang Heng went much further and constructed one that rotated by the force of flowing water in such a way that its movement coincided with the rising and setting of the stars. What is intriguing about this is whether he had devised some primitve form of clockwork — some early escapement, preparing the way for Su Sung's water clock of the 11th century — 1200 years before Giovanni de Dondi was to introduce it into Europe. Even more impressive is his construction of the world's first seismograph. It is clearly described as consisting of a vessel on the outside of which were eight dragon heads containing a ball. When an earthquake occurred the ball was propelled out of a dragon's mouth and caught by a bronze toad waiting underneath. Inside there was, presumably, some pendulum mechanism, which would release just one ball selectively giving the direction of the shock. It is interesting to note that the first seismograph recorded in the west, depending on the spilling of an overfilled saucer of mercury, was in 1703.

Chapman, Sydney. (b. Jan. 29, 1888; Eccles, England; d. June 16, 1970; Boulder, Colorado). British mathematician and geophysicist. Chapman entered Manchester University in 1904 to study engineering. After graduating in 1907, his interest was diverted into more strictly mathematical areas, and he went to Cambridge to study mathematics, graduating in 1910. His first post was as chief assistant at the Royal Observatory, Greenwich, and his work there sparked off his lasting interest in a number of fields of applied mathematics, notably geomagnetism. In 1914 Chapman returned to Cambridge as a lecturer in mathematics, and in 1919 he moved back to Manchester as professor of mathematics, remaining there for five years. From 1924 to 1946 he was professor of mathematics at Imperial College, London. After working at the War Office during World War II he moved to Oxford to take up the Sedleian Chair in natural philosophy, from which he retired in 1953. However, his retirement meant no lessening in his teaching and research activity, which continued for many years at the Geophysical Institute, Alaska, and at the High Altitude Observatory at Boulder, Colorado. The two main topics of Chapman's mathematical work were the kinetic theory of gases and geomagnetism. In the 19th century James Clerk Maxwell and Ludwig Boltzmann had put forward ideas about the properties of gases as determined by the motion of the molecules of the gas. Chapman's work, which he began in 1911, was the next major step in the development of a full mathematical treatment of the kinetic theory. The Swedish mathematician Enskog had been working, independently of Chapman, along similar lines, and the resulting theory is now generally known as the *Chapman - Enskog theory of gases.* While working in 1917 on mixtures of gases Chapman predicted the phenomenon of gaseous thermal diffusion. His subsequent work on the upper atmosphere was a practical application of his earlier more theoretical study of gases. Highlights of Chapman's work on geomagnetism are his work on the variations in the Earth's magnetic field in periods of a lunar day (27.3 days) and its submultiples. This he showed to be the result of a small tidal movement set up in the Earth's atmosphere by the Moon. He also developed, in 1930, in collaboration with one of his students, what has become known as the *Chapman-Ferraro theory* of magnetic storms. In collaboration with Julius Bartels, Chapman wrote *Geomagnetism* (2 vols. 1940), which soon established itself as a standard work.

Charles P. Steinmetz. Charles Proteus Steinmetz had a brilliant record in the fields both of mathematics and electrical engineering, yet his early ambition was not concerned with electricity. As far as he had any definite aspiration, he had set his heart upon becoming a professor of mathematics in his native country—Germany. When he came to America, after a sudden dramatic withdrawal from his homeland, circumstances more than anything else brought him into business relations with Rudolph Eickemeyer, of Yonkers. And there he became linked inseparably with electrical affairs and their problems.

Steinmetz was born on April 9, 1865, in Breslau, capital of the south German province of Silesia. His father was Carl Heinrich Steinmetz, an expert lithographer, employed at that time in the headquarters office of the Ober Schlesische Railroad. His mother died when he was a year old. The name given him at his christening was Carl August Rudolph Steinmetz, and under that name he began going to school at the age of five. He entered the University of Breslau, at the age of 17, and first attracted attention by attending every lecture in mathematics and astronomy during his first year. Thereafter he studied six subjects every year, yet joined heartily in all the student merry-makings as a member of the undergraduate mathematical society. When he joined the latter, his fellow students gave him the nickname of "Proteus". He not only worked hard at his studies but also affiliated himself with a small group of students who formed part of the German socialist movement. For a couple of years he took part in secret meetings, read socialist literature and even helped get out a socialist newspaper. But at length the activity of Bismarck's police resulted in the arrest of many of Steinmetz's friends, including several students, and his own apprehension. To escape a prison sentence, he hurriedly departed from Germany early in 1888, just as he was about to receive his university degree. He went secretly to Vienna and thence to Zurich, where he studied mechanical engineering for a year at the university there. Then, falling in with a young friend from America, he decided, in an adventurous moment, to accompany the latter to the United States. He had so little money, however, that he accepted the offer of his friend to pay his travelling expenses. And so, on June 1, 1889, Steinmetz arrived in New York having traveled in the steerage on a French immigrant steamer. He could speak scarcely any English, and he was destitute of funds. But for his friend's intercession, he would not have been allowed to land. After he did win an entrance into America, the young German spent two weeks tramping around New York and vicinity looking for work. Finally he went up to Yonkers to present a letter of introduction to Rudolph Eickemeyer. The latter eventually hired Steinmetz to be assistant draftsman at the Eickemeyer factory at twelve dollars a week. From that year Steinmetz became more and more active in electrical science, and electrical research. He and Eickemeyer together worked on various electrical problems until in 1890 Steinmetz began an investigation into the magnetic properties of iron. He sought to discover what nobody knew at the time, and what all electrical engineers desired to know — the law of hysteresis losses in alternating current apparatus. Taking all the data at hand, he performed some intricate mathematical calculations, which he verified with tests. The outcome of it all was the presentation by him on January 19 and September 27, 1892, of a memorable paper of nearly 200 pages in length, before the American Institute of Electrical Engineers, in which he defined and elucidated the law of hysteresis losses and thereby put an end forever to the hit-or-miss methods of designing alternating current machines which had been practiced up to that time. This paper immediately established his reputation. In 1892, the newly formed General Electric Company bought out Eickemeyer's business; and by the terms of that deal, Steinmetz's services were transferred to the new company. He went to Lynn and was placed in the calculating department. In the spring of 1894 he was transferred to Schenectady. He planned the design of the company's apparatus, continued his research and development work and finally organized the company's consulting engineering department, becoming himself chief consulting engineer. His funda mental investigations in electrical science continued. In 1894 he brought out his theory of alternating current calculation, which introduces the general number, and is in itself revolutionary. He then investigated transient electrical phenomena, including lightning, resulting in his explanation of the origin of lightning and in the design by his laboratory of a generator for reproducing lightning on a small scale, by means of which valuable improvements in the design of lightning arresters have become possible. These experiments of his created a great stir in the popular mind, which was fascinated by the notion of a mere human being successfully imitating nature's destructive bolt. Steinmetz learned the language in less than a year after his arrival, observed and adopted our customs and took steps without delay for securing his naturalization papers. Within five years after landing he became a citizen of the United States. Seeking to Americanize even his name, he changed the Carl to the English form of Charles, and substituted for his two long German middle names the old university nickname of Proteus. He interested himself intelligently in the affairs of his own community, and twice served as president of the Schenectady Board of Education, being appointed first in 1911 by Mayor George R. Lunn, who was then a socialist. In 1915, when the latter ran again on a socialist ticket, Dr. Steinmetz was elected president of the Common Council. As a member of the Board of Education for some years, his work for the public schools was constructive and directed entirely towards the public welfare. Harvard University, in 1902, conferred upon Dr. Steinmetz the honorary degree of M.A.; and in the following year he received the honorary degree of Ph.D. from Union College. In the latter institution he served a number of years on the faculty as professor of electrical engineering, and later as professor of electro-physics. His technical works are many, including the results of his three major investigations, based on the original papers: on the magnetic circuit and the law of hysteresis; into the theory and calculation of alternating current phenomena; and concerning transient phenomena. Most of his papers on electrical subjects appeared in the transactions of the American Institute of Electrical Engineers, of which, at one time, he was president. He was also president of the Illuminating Engineering Society, honorary member of the National Electric Light Association, a fellow of the American Association for the Advancement of Science, a member of the American Mathematical Society, the Society of Mechanical Engineers, the Electrochemical Society, the British Institute of Electrical Engineers, the German Electrochemical Society and other organizations. His later contributions to his profession were mainly in the realm of hydro-electric science, in the future development of which and what it will do for America, he was an enthusiastic believer. Dr. Steinmetz died at Schenectady on October 26, 1923.

Chou Kung. (fl. 12th century BC). Chinese mathematician. Chou Kung, or the duke of Chou, was the brother of Wu Wang, the founder of the Chou dynasty. He served briefly as regent on his brother's death. He is remembered for his name in the *Chou-li,* one of the earliest Chinese mathematical works, in which he supposedly takes part in a dialog with someone called Shang Kao. The dialog was thought to date back to the time of Chou Kung but scholars now think this exravagant. Although they are prepared to accept some parts as going back to the sixth century BC, the bulk of it they assign to the Han dynasty (200 BC - AD 200). The work is important in providing hard evidence for the state of early Chinese mathematics. The most significant feature of the work is a demonstration of the truth of Pythagoras's theorem for triangles with sides of 3, 4, and 5 units. The 'proof', described as 'piling up the rectangles', is purely diagrammatic. The work also shows knowledge of the multiplication and division of fractions, the finding of common denominators, and the extraction of square roots. Compared with Greek works of a comparable period, such as Archimedes', the work is unimpressive and gives little indication of future achievement.

Christian Huygens. After the death of Galileo and of Kepler, and before Sir Isaac Newton rose to fame, there is an interval during which Huygens stands out as by far the greatest astronomer of his day. Born at the Hague on April 14, 1629, Christian Huygens de Zuylichem belonged to a wealthy and distinguished family. His father was secretary and adviser to three successive princes of Orange, and his elder brother, Constantine, succeeded to this responsible office and went to England in 1688 with William of Orange. Christian showed early signs of a very remarkable mind and his father, as was usual in those days, educated him as rapidly and widely as possible. He was instructed in music as well as in the literary subjects which were customary at the period, and at the age of thirteen he was introduced to the construction of machines, an occupation for which he gave evidence of extraordinary ability. At sixteen years he was sent to Leyden to study law, and twelve months later to the University of Breda, where he plunged into mathematics, which always remained his favorite study. His first mathematical essays brought him to the notice of Descartes. On leaving the university, in 1649, he traveled with the Count of Nassau, and on his return to Holland set to work with enthusiasm on the most abstruse mathematical problems and on the mechanical inventions which are his chief title to fame. He published treatises on the quantities of the parabola and other geometrical figures, and at the same time began to grind and polish lenses for large telescopes, becoming the maker of instruments superior to any which existed at the time. It was with a twelve-foot telescope of his own handiwork that he not only discovered, in 1656, a sixth satellite of Saturn, Titan, but determined the period of its revolution. Huygens was the first to apply the pendulum to clocks, and to employ the device to determine the acceleration of gravity. It would be impossible even to enumerate the many astronomical discoveries which Huygens initiated, but certainly his most illustrious observations were those by which he first understood the nature of Saturn's rings. That planet had been a most baffling puzzle to astronomers ever since the telescope had revealed its apparently triple form, for Saturn changes remarkably in appearance according as the rings are viewed edgewise — when they disappear, except to the most powerful telescopes — or are seen more or less in face. Having by far the best optical instrument of his age, Huygens was able, in 1656, to announce the true relation of Saturn and its rings, though he was not able to separate the ring-system into its component members. In "The System of Saturn", wherein the whole matter, so far as at that time disclosed, was clearly set forth, he records other important astronomical observations. For example, he describes the bands on Jupiter, the principal markings of Mars and the Orion nebula. He was able also to define the very significant fact that, even in the largest telescopes, the stars, as distinguished from planets, have no diameter, but are more points of light without dimensions. From this discovery he very naturally turned to the determination of the diameters of the several planets, and invented an instrument for the purpose. This, as modified later, became the micrometer which is used in modern observatories. It should be said, however, that a previous worker in the same field, Gascoigne, who was killed at the battle of Marston Moor in 1644, had constructed a true flair micrometer, but the invention remained unknown for many years afterward. Huygens visited England in 1660, and was admitted a member of the Royal Society. It was a time of general scientific revival, and the French Academy of Sciences, which was being founded about the same time, persuaded Huygens to take part in its direction, offering him at the same time princely remuneration. He acceded to their request, and remained in Paris until the year 1681, when he resigned his position on account of failing health, and returned to his native Netherlands. To a man of this type, Sir Isaac Newton's *"Principia"*, which he secured in 1689, was a revelation and a delight. His first impulse was to cross the Channel once more, and get to know the author and, having done so, he wrote and published two important works, under the influence of Newton, on light and on weight respectively. But he was now coming near to the end of his days. His last illness began early in 1695, and he died on July 8 of the same year. The work he did was much more copious than can be even suggested by this brief notice. He did important work in optics and he is regarded as the founder of the waves theory of light. He had few equals in higher mathematics, and Newton praised his style and methods of mathematical exposition, and paid him the compliment of confessed imitation. He was a great expounder of the differential calculus, initiated by Pascal but still at this time generally unknown. Another important work dealt with the movements resulting from percussion, being an inquiry closely related to the modern science of ballistics. A lighter and more fanciful late work, entitled "*Cosmotheoros*", consists of conjectures on the physical constitution of distant worlds and their inhabitants.

Chu Shih-Chieh. (fl. 1300). Chinese mathematician. Little seems to be known about the life of Chu Shih-Chieh other than that he traveled around China teaching mathematics and that he wrote two important works, the *Introduction to Mathematical Studies* (1299) and, one of the high-water marks of Chinese mathematics, the *Precious Mirror of the Four Elements* (1303). The first work states clearly the rule of signs for both algebraic addition and multiplication. Its treatment of negative signs, elementary as it may now seem, contrasts markedly with the practice of the ancients and the medieval West in ignoring them. The second work contains an advanced treatment of equations both simple and higher. It begins with a remarkedly clear diagram of what came to be known as Pascal's triangle. He also produced the first formula in China for the sum of arithmetical progressions.

Clairaut, Alexis-Claude. (b. May 7, 1713; Paris; d. May 17, 1765; Paris). French mathematician and physicist. Clairaut contributed to many fields of science, chief among these being mathematics, physics, and geodesy. He was a child prodigy and was elected to the Paris

Academy of Sciences for promising work in geometry when he was only 18. After joining the academy Clairaut developed a lively interest in geodesy. He was an early supporter of Newton at a time when his views were not yet popular in France. Clairaut is known to have studied mathematics briefly with Johann I Bernoulli in Basel. In 1736 Clairaut's interest in geodesy took a practical turn when he embarked on an expedition to Lapland led by Pierre Maupertuis. This was undertaken in order to measure a meridian arc of one degree inside the Arctic Circle. This expedition successfully completed (1737), Clairaut's interests reverted to more mathematical matters and in particular to what was then known as 'celestial mechanics', as pioneered by Newton. Clairaut's enthusiasm for the ideas of Newton led him to collaborate on preparing a French translation of Newton's great work, the *Principia.* Clairaut also added a considerable amount of explanatory material of his own to clarify Newtons ideas and much of his best work found its way into these often extensive explanatory additions. This translation was a great influence in popularizing Newton's ideas in France. Clairaut's best and most important writings are devoted chiefly to mathematical astronomy and to algebra and geometry. He made valuable contributions to the development of the calculus and to the study of differential equations. He also did notable work in practical optics, which he put to good use in designing better telescopes, and was particularly interested in Newton's corpuscular theory of light. Clairaut wrote influential treatises on algebra and geometry.

Claudius Ptolemy. Claudius Ptolemaeus, generally known by the Anglicized form of his name, Ptolemy, lived in

Alexandria, Egypt, in the second century A.D. From 127 to 141 he was active as an astronomer, gathering the mathematical data that served as the basis for the most detailed geocentric cosmology ever formulated. The Ptolemaic version of the Universe prevailed virtually without question from the early Christian era to the high Renaissance. Ptolemy's cosmology was prompted by his conviction that the Universe must be spherical because only the sphere is perfect. He decided that stars must also move in circles for the same reason. Ptolemy's "symmetry principle" was to cause him many headaches as his observations revealed that the sun and planets depart conspicuously from uniform circular motion. As some of these bodies seemed to reverse their direction of motion periodically, Ptolemy assumed that they must move on circles within circles. The curve traversed by a point on a circle rolling in a circle is called an "epicycle." Ptolemy stripped these epicycles of the little physical significance they had to begin with and treated them as purely mathematical terms that could fit the measured changes in position and brightness of the renegade planets. Yet it often required more than one of these terms to explain these movements, so that Ptolemy's geocentric model gradually became very bulky and unworkable.Ptolemy's scientific works have come down to us in the *Almagest* and the *Geography. Almagest* is a hybrid Greek-Arabic title meaning "The Greatest [Compilation]"; the work is also known as the *Syntaxis*. It is a continuation and a vast amplification of the early work of Hipparchus, in which Ptolemy's greatest contribution is his detailed mathematical theory of the motions of the sun, moon, and planets: [The *Almagest*] ... cover[s] the whole of mathematical astronomy as the ancients conceived it. Ptolemy assumes in the reader nothing beyond a knowledge of Euclidean geometry and an understanding of common astronomical terms; starting from first principles, he guides [the reader] through the prerequisite cosmological and mathematical apparatus to an exposition of the theory of the motion of the heavenly bodies which the ancients knew (sun, moon, Mercury, Venus, Mars, Jupiter, Saturn, and the fixed stars, the latter being considered to lie on a single sphere concentric with the earth).... For each body in turn Ptolemy describes the type of phenomena that have to be accounted for, proposes an appropriate geometric model, derives the numerical parameters from selected observations, and finally constructs tables enabling one to determine the motion or phenomena in question for a given date. Ptolemy's work provided the first systematic classification of ancient Greek astronomy. The shortcomings of the geocentric theory to which he clung so tightly should not obscure the fact that Ptolemy's *Almagest* represents one of the supreme intellectual achievements in the history of science. Although its shortcomings have been amply documented through the ages — some critics going so far as to call Ptolemy a charlatan — few would dispute its clarity as well as the magnificent scope of its coverage.

Cocker, Edward. (b. 1631; d. 1675; London). English engraver and mathematician. Cocker was famous for writing and engraving a very influential and popular textbook, the *Arithmetic* (1678). He was also a teacher of arithmetic and writing. Cocker produced notable textbooks on other subjects, including several writing manuals and an English dictionary but none of these rivaled in popularity his book on arithmetic, which went through over 100 editions. Cocker is mentioned by Samuel Pepys who thought his skill as an engraver sufficient to comment favorably on it in his *Diary*.

Conon of Samos. (fl. 245 BC; Alexandria, Egypt). Greek mathematician and astronomer. Canon settled in Alexandria and was employed as court astronomer to the Egyptian monarch Ptolemy III. None of Conon's own writings survive and what is known of his work is through secondhand references to him by other Greek mathematicians. For example, Conon's work on conics was made use of by Apollonius of Perga in his famous treatise on conics. Among Conon's activities as an astronomer was the compilation of tables of the times of the rising and setting of the stars, known as the *parapegma*. He was also responsible for naming a constellation of stars. The consort of Ptolemy III, Berenice II, presented her hair as an offering at the temple of Aphrodite. This disappeared and Conon claimed that the hair now hung as a new constellation of stars, which he named Coma Berenices ('Berenice's Hair'). Conon was known to have been a friend of Archimedes and it is probable that the 'Spiral of Archimedes', a mathematical curve, was in fact Conon's discovery.

Coulson, Charles Alfred. (b. Dec. 13, 1910; Dudley, England; d. Jan. 8, 1974; Oxford, England). British theoretical chemist, physicist, and mathematician. Coulson's father was principal of a local technical college. He was educated at Cambridge University, where he obtained his PhD in 1935 and afterward taught mathematics at the universities of Dundee and Oxford. Coulson then successively held appointments as professor of theoretical physics at King's College, London (1947-52), Rouse Ball Professor of Mathematics at Oxford (1952-72), and, still at Oxford, professor of theoretical chemistry from 1972 until his death in 1974. As a physicist Coulson wrote the widely read *Waves* (1941) and *Electricity* (1948). His most creative work however was as a theoretical chemist. In 1933 he did early work on calculating the energy levels in polyatomic molecules and in 1937 he provided a theory of partial bond order (e.g. chemical bonding intermediate between double and single bonding). He also worked on aromatic molecules (benzene, naphthalene, etc.) with Christopher Longuet-Higgins. His book *Valence* (1952) covers the application of quantum mechanics to chemical bonding. Later Coulson turned to theoretical studies of carcinogens, drugs, and other topics of biological interest. Coulson was also one of the leading Methodists of his generation and, from 1965 to 1971, chairman of Oxfam, the third-world charity. He produced a number of works on the relationship between Christianity and science of which *Science and Christian Belief* (1955) is a typical example.

Courant, Richard. (b. Lublinitz, Silesia, Germany [now Poland], 8.1.1888; naturalized American citizen, 1940; d. New Rochelle, NY, USA, 27.1.1972). German/American mathematician, also important in the development of the mathematics profession. Educated at Göttingen, he worked in differential equations and their applications to physical problems. *Methoden der mathematischen Physik* (2 vols, Berlin, 1931; *Methods of Mathematical Physics,* NY, 1953), written by him but to which he added Hilbert's name, turned out to be a timely primer of mathematical techniques for quantum mechanics. During the 1920s he also inaugurated a Mathematical Institute in Göttingen, and launched important journals and a research monograph series (the 'Yellow Books') with Springer-Verlag. As a Jew, Courant was suspended from his appointment in 1931. An uncertain period ended with an offer in 1934 from New York University, where he spent the rest of his career. He built up its research programme substantially, culminating in the founding of the Courant Institute of Mathematical Studies in 1965. During his New York career he also published another volume of *Methods of Mathematical Physics* (NY, 1937), research monographs on specific topics (including shock waves), and a popular book called *What is Mathematics?* (with H. Robbins, Oxford, 1941).

D

D'Alembert, Jean Le Rond. (b. Nov. 16, 1717; Paris; d. Oct. 29, 1783; Paris). French mathematician, encyclopedist, and philosopher. D'Alembert was the illegitimate son of a Parisian society hostess, Mme de Tenzin, and was abandoned on the steps of a Paris church, from which he was named. He was brought up by a glazier and his wife, and his father, the chevalier Destouches, made sufficient money available to ensure that d'Alembert received a good education although he never acknowledged that d'Alembert was his son. He graduated from Mazarin College in 1735 and was admitted to the Academy of Sciences in 1741. D'Alembert's mathematical work was chiefly in various fields of applied mathematics, in particular dynamics. In 1743 he published his *Traité de dynamique,* in which the famous *d'Alembert principle* is enunciated. This principle is a generalization of Newton's third law of motion, and it states that Newton's law holds not only for fixed bodies but also for those that are free to move. D'Alembert wrote numerous other mathematical works on such subjects as fluid dynamics, the theory of winds, and the properties of vibrating strings. His most significant purely mathematical innovation was his invention and development of the theory of partial differential equations. Between 1761 and 1780 he published eight volumes of mathematical studies. Apart from his mathematical work he is perhaps more widely known for his work on Denis Diderot's *Encyclopédie* as editor of the mathematical and scientific articles, and his association with the philosophes. D'Alembert was a friend of Voltaire's and he had a lively interest in theater and music, which led him to conduct experiments on the properties of sound and to write a number of theoretical treatises on such matters as harmony. He was elected to the French Academy in 1754 and became its permanent secretary in 1772 but he refused the presidency of the Berlin Academy.

Daniel Bernoulli. When you read a reference to a Bernoulli in a book dealing with physics or mathematics, you have to make sure you know which Bernoulli is involved. In science the Bernoullis are rather like the Bach family in music. In the course of a century, eight members pursued mathematical studies, and several made very substantial contributions to the sciences and related disciplines. According to E.T. Bell, "[n]o fewer than 120 of the descendants of the mathematical Bernoullis have been traced genealogically, and of this considerably posterity the majority achieved distinction - sometimes amounting to eminence — in the law, scholarship, science, literature, the learned professions, administration, and the arts. None were failures." They were Flemish by origin, but, being Protestants, they were driven out of the Netherlands in the sixteenth century and moved to Switzerland. Jacob Bernoulli was a contemporary of Newton and Leibniz and was decidedly in their class of mathematical skill. His brother, Johann Bernoulli, was also a gifted mathematician. Johann was teaching in Groningen in the Netherlands when his son Daniel was born in 1700. The family returned to Switzerland when Daniel was five years old. It is surprising that Johann Bernoulli did everything he could to dissuade young Daniel from studying mathematics. Daniel was taught geometry by his brother Nicolaus. He studied philosophy and logic, earning his baccalaureate in 1715 and his master's degree one year later. His failure as a commercial apprentice led him to study medicine at Basel, Heidelberg and Strasbourg. He returned to Basel after completing his doctorate in 1721. Although unsuccessful in securing vacant chairs at Basel in anatomy and biology, Bernoulli's 1724 publication of his *Exercitationes mathematicae* brought him fame and an appointment as professor of mathematics at St. Petersburg. He returned to Basel eight years later to become professor of anatomy and botany. Yet he did not obtain the chair in physics at Basel until 1750. He won many prestigious mathematics honors, including ten prizes awarded at various times by the French Academy. His major contributions were to probability theory, calculus, mechanics, and differential equations. Daniel Bernoulli's most famous book is the *Hydrodynamica,* which represented both a culmination and a synthesis of many of his ideas: [In the *Hydrodynamica* Bernoulli] laid the foundations, theoretical and practical, for the "equilibrium, pressure, reaction and varied velocities" of fluids. The *Hydrodynamica* is notable also for presenting the first formulation of the kinetic theory of gases, a keystone of modern physics. Bernoulli showed that, if a gas he imagined to consist of "very minute corpuscles," "practically infinite in number," "driven hither and thither with a very rapid motion," their myriad collisions with one another and impact on the walls of the containing vessel would explain the phenomenon of pressure. Moreover, if the volume of the container were slowly decreased by sliding in one end like a piston, the gas would be compressed, the number of collisions of the corpuscles would be increased per unit of time, and the pressure would rise. The same effect would follow from heating the gas; heat, as Bernoulli perceived, being nothing more than "an increasing internal motion of the particles." This "astonishing prevision of a state of physics which was not actually reached for 110 years" (notably by Joule, who calculated the statistical averages of the enormous number of molecular collisions and thus derived Boyle's law — pressure x volume = constant — from the laws of impact) was fully sustained by Bernoulli's remarkable experimental and theoretical labors.... The Bernoulli paper offered here covers the above topics and is excerpted from the tenth section of the *Hydrodynamica*. It is one of the earliest versions of the kinetic theory of gases, dating from 1738. As noted above, the work is well ahead of its time; the full mathematical development of Bernoulli's thought had to await the arrival of James Clerk Maxwell and Ludwig Boltzmann.

Darwin, Sir George Howard. (b. July 9, 1845; Down, England; d. Dec. 7, 1912; Cambridge, England). British astronomer and mathematician. Darwin was the second son of the famous biologist, Charles Darwin. He was educated at Clapham Grammar School, where the astronomer Charles Pritchard was headmaster, and Cambridge University. He became a fellow in 1868 and, in 1883, Plumian Professor of Astronomy, a post he held until his death. He was knighted in 1905. His most significant work was on the evolution of the Earth-Moon system. His basic premise was that the effect of the tides has been to slow the Earth's rotation thus lengthening the day and to cause the Moon to recede from the Earth. He gave a mathematical analysis of the consequences of this, extrapolating into both the future and the past. He argued that some 4.5 billion years ago the Moon and the Earth would have been very close, with a day being less than five hours. Before this time the two bodies would actually have been one, with the Moon residing in what is now the Pacific Ocean. The Moon would have been torn away from the Earth by powerful solar tides that would have deformed the Earth every 2.5 hours. Darwin's theory, worked out in collaboration with Osmond Fisher in 1879, explains both the low density of the Moon as being a part of the Earth's mantle, and also the absence of a granite layer on the Pacific floor. However, the theory is not widely accepted by astronomers. It runs against the Roche limit, which claims that no satellite can come closer than 2.44 times the planet's radius without breaking up; there are also problems with angular momentum. Astronomers today favor the view that the Moon has formed by processes of condensation and accretion. Whatever its faults, Darwin's theory is important as being the first real attempt to work out a cosmology on the principles of mathematical physics.

Dedekind, (Julius Wilhelm) Richard. (b. Oct. 6, 1831; Brunswick, now in West Germany; d. Feb. 12, 1916; Brunswick). German mathematician. Dedekind's life was long but outwardly uneventful. He studied in Brunswick (Braunschweig) at the Caroline College and then at the University of Göttingen where his teachers included Karl Gauss. He taught briefly at Göttingen and at the Federal Institute of Technology, Zurich, and then returned to Brunswick to take up a post at the Technical Institute. Dedekind remained in this relatively obscure position for nearly fifty years. His most important contribution to mathematics was in his work on the foundations of the theory of numbers. The axioms for the natural number system, which are now known as the *Peano axioms,* were in fact first formulated by Dedekind in his seminal essay *Was sind und was sollen die Zahlen* (1888; What Numbers Are and Should Be). His best-known contribution was the rigorous definition of irrational numbers as classes of fractions by means of *Dedekind cuts* (or *sections*). Dedekind's work marks an extremely important step forward in the movement in 19th-century mathematics toward modern standards of rigor. Dedekind was a close friend of Georg Cantor and one of the few mathematicians to recognize immediately the value of Cantor's work on the transfinite. He defined the real numbers by means of the DEDEKIND CUT, and originated the concepts of RING (in the sense of DEDEKIND RING) and UNIT and the definition of IDEAL in algebra.

de Finetti, Bruno. (b. Innsbruck, Tyrol, Austria-Hungary [now Austria], 13.6.1906). Italian statistical theorist, working at the University of Rome. He developed subjective probabilities, given not only to events but to hypotheses, when the calculus of probabilities becomes a normative theory for testing ideas against evidence. Ideas are to be arranged so that all relevant hypotheses can be listed and do not interact, with their probabilities mutually comparable and consistent; there is then an unequivocal rule for adjusting their probabilities to take account of new evidence. Spurred by the parallel thinking of L.J. Savage and the mediation of D.V. Lindley, this has developed over 50 years into a substantial alternative theory of statistics which sometimes reinterprets and sometimes replaces classical methods.

Delambre, Jean Baptiste Joseph. (b. Sept. 19, 1749; Amiens, France; d. Aug. 19, 1822; Paris). French astronomer and mathematician. Delambre was most unusual for a mathematician and astronomer in that he did not begin the serious study of his subject until he was well over 30 years old. As a student he had been interested in the classics and only turned to the exact sciences when he was 36. He published tables of Jupiter and Saturn in 1789 and of Uranus in 1792. He also measured an arc of the meridian between Dunkirk and Barcelona to establish a basis for the new metric system. He succeeded Joseph de Lalande as professor of astronomy at the Collège de France in 1795. In his later years he devoted himself to a monumental six-volume *Histoire de l'astronomie* (1817-27).

De La Rue, Warren. (b. Jan. 18, 1815; Guernsey, Channel Islands; d. Apr. 19, 1889; London). British astronomer, and mathematician. De la Rue was the son of a printer and worked most of his life in his father's business. He was educated in Paris and studied science privately. He was initially interested in chemistry, being a friend of and working with August Hofmann, but later, at the suggestion of James Nasmyth, he took up astronomy, building a small observatory for himself. De la Rue devoted himself to problems of photographic astronomy. He was the first to apply the collodion process (invented by Frederick Archer in 1851) to photographing the Moon. In 1852, he took some photographs that were sharper than any previously produced and that could be enlarged without blurring. Ten years later he was producing photographs that could show as much as could be seen through any telescope. In 1854 he designed the photo-heliograph, a device for taking telescopic photographs of the Sun. In 1860 he used it to take dramatic photographs of prominences during the total eclipse in Spain, proving that they were solar (and not, as had been thought, lunar) in origin. De la Rue gave up active astronomical investigation in 1873 donating his telescope to the observatory at Oxford and devoting the rest of his life to his business and to his chemical researches.

De Moivre, Abraham. (b. May 26, 1667; Vitry, France; d. Nov. 27, 1754; London). French mathematician. Although born in France De Moivre was a Huguenot and consequently was forced to flee to England to escape the religious persecution that flared up in 1685 after the revocation of the Edict of Nantes. In England he came to know both Isaac Newton and Edmond Halley, eventually becoming a fellow of the Royal Society of London himself in 1697. De Moivre made important contributions to mathematics in the fields of probability and trigonometry. His interest in probability was no doubt stimulated by the fact that despite his abilities he was unable to find a permanent post as a mathematician and so was forced to earn his living by, among other things, gambling. De Moivre was the first to define the concept of statistical independence and to introduce analytical techniques into probability. His work on this was published in *The Doctrine of Chances* (1718), later followed by *Miscellanea annalytica* (1730; Analytical Miscellany). De Moivre also introduced the use of complex numbers into trigonometry. *De Moivre's theorem* is the relationship $(\cos A + i \sin A)^n = \cos nA + i \sin nA$.

Desargues, Girard. (b. Mar. 2, 1591; Lyons, France; d. October 1661; France). French mathematician and engineer. Little is known of Desargues' early life. He did serve as an engineer at the siege of La Rochelle (1628) and later became a technical adviser to Cardinal de Richelieu and the French government. He is said to have known René Descartes. Around 1630 Desargues joined a group of mathematicians in Paris and concentrated on geometry. In his most famous work, *Brouillon projet d'une atteinte aux événemens des rencontres d'une cône avec un plan* (1639; Proposed Draft of an Attempt to deal with the Events of the Meeting of a Cone with a Plane), he applied projective geometry to conic sections. *Desargues' theorem* states that if the corresponding points of two triangles in nonparallel planes in space are joined by three lines that intersect at a single point, then the pairs of lines that are the extensions of corresponding sides will each intersect on the same line. Blaise Pascal was greatly influenced by Desargues, whose contribution to projective geometry was not recognized until a handwritten copy of his work was found in 1845. This oversight probably arose because he used obscure botanical symbols instead of the better-known Cartesian symbolism.

Descartes, René du Perron. (b. Mar. 31, 1596; La Haye, France; d. Feb. 11, 1650; Stockholm, Sweden). French mathematician, philosopher, and scientist. Descartes' father was a counselor of the Brittany *parlement* while his mother, who died shortly after his birth, left him sufficient funds to make him financially independent. He was educated by the Jesuits of La Fléche (1604-12) and at the University of Poitiers, where he graduated in law in 1616. For the next decade Descartes spent much of his time in travel throughout Europe and in military service, first with the army of the Prince of Orange,

Maurice of Nassau, and later with the Duke of Bavaria, Maximilian, with whom he was present at the battle of the White Mountain outside Prague in 1620. In the years 1628-49 Descartes settled in the freer atmosphere of Holland. There, living quietly, he worked on the exposition and development of his system. Somewhat unwisely, he allowed himself to be enticed into the personal service of Queen Christina of Sweden in Stockholm in 1649. Forced to indulge the Queen's passion for philosophy by holding tutorials with her at 5 am on icy Swedish mornings Descartes, who normally loved to lie thinking in a warm bed, died within a year from pneumonia and the copious bleeding inflicted by the enthusiastic Swedish doctors. Descartes is in many ways, in mathematics, philosophy, and science, the first of the moderns. The moment of modernity can be dated precisely to 10 November, 1619, when, as later described in his *Discours de la méthode* (1637; Discourse on Method), he spent the whole day in seclusion in a *poéle* (an overheated room). He began systematically to doubt all supposed knowledge and resolved to accept only "what was presented to my mind so clearly and distinctly as to exclude all ground of doubt." Descartes thus managed to pose in a single night the problem whose solution would obsess philosophers for the next 300 years. The same night also provided him with one of the basic insights of modern mathematics — that the position of a point can be uniquely defined by coordinates locating its distance from a fixed point in the direction of two or more straight lines. This was revealed in his *La Geométrie* (1637), published as an appendix to his *Discours,* and describing the invention of analytic or coordinate geometry, by which the geometric properties of curves and figures could be written as and investigated by algebraic equations. The system is known as a *Cartesian coordinate* system. His theories on physics were published in his *Principia philosophiae* (1644; Principles of Philosophy). "Give me matter and motion and I will construct the universe," Descartes had proclaimed. The difficulty for him arose from his account of matter which, on metaphysical grounds, he argued, "does not at all consist in hardness, or gravity or color or that which is sensible in another manner, but alone in length, width, and depth," or, in other words, extension. From this initial handicap Descartes was forced to deny the existence of the void and face such apparently intractable problems as how bodies of the same extension could possess different weights. With such restrictions he was led to describe the universe as a system of vortices. Matter came in three forms — ordinary matter opaque to light, the ether of the heavens transmitting light, and the subtle particles of light itself. With considerable ingenuity and precious little concern for reality Descartes used such a framework within which he was able to deal with the basic phenomena of light, heat, and motion. Despite its initial difficulties it was developed by a generation of Cartesian disciples to pose as a viable alternative to the mechanics worked out later in the century by Newton. Unlike many less radical thinkers Descartes did not shrink from applying his mechanical principles to physiology, seeing the human body purely in terms of a physico-mechanical system with the mind as a separate entity interacting with the body via the pineal gland — the supposed seat of the soul. The fundamental impact of Descartes' work was basically one of demystification. Apart from the residual enigma of the precise relationship between mind and body, the main areas of physics and physiology had been swept clear of such talk as that of occult powers and hidden forms. Descartes founded ANALYTIC GEOMETRY, and introduced EXPONENTIAL NOTATION, CARTESIAN COORDINATES, and methods of solving POLYNOMIAL EQUATIONS to mathematics. His work as a whole was ruled by the desire to systematize all knowledge as resting only on what is clearly self-evident, on the axiomatic model of Euclid's geometry, and thereby to achieve certainty. His method involved the suspension of belief in anything that could conceivably be doubted, and he then founded his edifice on the argument that, while doubting, one cannot doubt that one doubts, and therefore thinks; he expressed this single self-justifying proposition in his famous Latin phrase 'Cogito, ergo sum' ('I think, therefore I am'). The adjective *Cartesian* is derived from the archaic spelling *Des Cartes*.

De Sitter, Willem. (b. May 6, 1872; Sneek, Netherlands; d. Nov. 20, 1934; Leiden, Netherlands). Dutch astronomer and mathematician. De Sitter, the son of a judge, studied mathematics and physics at the University of Groningen, his interest in astronomy being aroused by Jacobus Kapteyn. After serving at the Cape Town Observatory from 1897 to 1899 and, back at Groningen, as assistant to Kapteyn from 1899 to 1908, he was appointed to the chair of astronomy at the University of Leiden. He also served as director of Leiden Observatory from 1919 to 1934. De Sitter is remembered for his proposal in 1917 of what came to be called the 'de Sitter universe' in contrast to the 'Einstein universe'. Einstein had solved the cosmological equations of his general relativity theory by the introduction of the cosmological constant, which yielded a static universe. But de Sitter, in 1917, showed that there was another solution to the equations that produced a static universe if no matter was present. The contrast was summarized in the statement that Einstein's universe contained matter but no motion while de Sitter's involved motion without matter. The Russian mathematician Alexander Friedmann in 1922 and the Belgian George Lemaître independently in 1927 introduced the idea of an expanding universe that contained moving matter. It was then shown in 1928 that the de Sitter universe could be transformed mathematically into an expanding universe. This model, the 'Einstein-de Sitter universe', comprised normal Euclidean space and was a simpler version of the Friedmann-Lemaître models in which space was curved. De Sitter also worked on celestial mechanics and stellar photometry. He spent much time trying to calculate the mass of Jupiter's satellites from the small perturbations in their orbits. The results were published in 1925 in his *New Mathematical Theory of Jupiter's Satellites.*

Diophantus of Alexandria. (fl. 250). Greek mathematician. Diophantus was one of the outstanding mathematicians of his era but almost nothing is known of his life and his writings survive only in fragmentary form. His most famous work was in the field of number theory and of the so-called *Diophantine equations* named after him. His major work, the *Arithmetica,* contained many new methods and results in this field. It originally consisted of 13 books but only 6 of these survived to be translated by the Arabs. However Diophantus was not solely interested in equations with only integral (whole number) solutions and also considered rational solutions. Diophantus made considerable innovations in the use of symbolism in Greek mathematics — the lack of suitable symbolism had previously hampered work in algebra.

Dirac, Paul Adrien Maurice. (b. Aug. 8, 1902; Bristol, England). British mathematician and physicist. Dirac's father was Swiss by birth. After graduating in 1921 in electrical engineering at Bristol University, Dirac went on to read mathematics at Cambridge University, where he obtained his PhD in 1926. After several years spent lecturing in America, he was appointed (1932) to the Lucasian Professorship of Mathematics at Cambridge, a post he held until his retirement in 1969. In 1971 he became professor of physics at Florida State University. Dirac is acknowledged as one of the most creative of the theoreticians of the early 20th century. In 1926, slightly later than Max Born and Pascual Jordan in Germany, he developed a general formalism for quantum mechanics. In 1928 he produced his relativistic theory to describe the properties of the electron. The wave equations developed by Erwin Schrödinger to describe the behavior of electrons were nonrelativistic. A significant deficiency in the Schrödinger equation was its failure to account for the electron spin discovered in 1925 by Samuel Goudsmit and George Uhlenbeck. Dirac's rewriting of the equations to incorporate relativity had considerable value for it not only predicted the correct energy levels of the hydrogen atom but also revealed that some of those levels were no longer single but could be split into two. It is just such a splitting of spectral lines that is characteristic of a spinning electron. Dirac also predicted from these equations that there must be states of negative energy for the electron. In 1930 he proposed a theory to account for this that was soon to receive dramatic confirmation. He began by taking negative energy states to refer to those energy states below the lowest positive energy state, the ground state. If there were a lower energy state for the electron below the ground state then, the question arises, why do some electrons not fall into it? Dirac's answer was that such states have already been filled with other electrons and he conjured up a picture in which space is not really empty but full of particles of negative energy. If one of these particles were to collide with a sufficiently energetic photon it would acquire positive energy and be observable as a normal electron, apparently appearing from nowhere. But it would not appear alone for it would leave behind an empty hole, which was really an absence of a negatively charged particle or, in other words, the presence of a positively charged particle. Further, if the electron were to fall back into the empty hole it would once more disappear, appearing to be annihilated together with the positively charged particle, or positron as it was later called. Out of this theory there emerged three predictions. Firstly, that there was a positively charged electron, secondly, that it could only appear in conjunction with a normal electron, and, finally, that a collision between them resulted in their total common annihilation. Such predictions were soon confirmed following the discovery of the positron by Carl Anderson in 1932. Dirac had in fact added a new dimension of matter to the universe, namely antimatter. It was soon appreciated that Dirac's argument was sufficiently general to apply to all particles. In 1937 Dirac published a paper entitled *The Cosmological Constants* in which he considered 'large-number coincidences'; i.e. certain relationships that appear to exist between the numerical properties of some natural constants. An example is to compare the force of electrostatic attraction between an electron and a proton with the gravitational attraction due to their masses. The ratio of these is about 10^{40}:1. Similarly, it is also found that the characteristic 'radius' of the universe is 10^{40} times as large as the characteristic radius of an electron. Moreover, 10^{40} is approximately the square root of the number of particles in the universe. These coincidences are remarkable and many physicists have speculated that these apparently unrelated things may be connected in some way. The ratios were first considered in the 1930s by Arthur Eddington, who believed that he could calculate such constants and that they arose from the way in which physics observes and interprets nature. Dirac used the 10^{40} number above in a model of the universe. He argued that there was a connection between the force ratio and the radius ratio. Since the radius of the universe increased with age the gravitational 'constant', on which the force ratio depends, may decrease with time. Above all else however Dirac was a quantum theorist. In 1930 he published the first edition of his classic work *The Principles of Quantum Mechanics.* In 1933 he shared the Nobel Prize for physics with Schrödinger.

Dirichlet, (Peter Gustav) Lejeune. (b. Feb. 13, 1805; Düren, now in West Germany; d. May 5, 1859; Göttingen, now in West Germany). German mathematician.Dirichlet studied mathematics at Göttingen where he was a pupil of Karl Gauss and Karl Jacobi. He also studied briefly in Paris where he met Joseph Fourier, who stimulated his interest in trigonometric series. In 1826 he returned to Germany and taught at Breslau and later at the Military Academy in Berlin. He then moved to the University of Berlin, which he only left 27 years later when he returned to Göttingen to fill the chair left vacant by Gauss's death. Dirichlet's work in number theory was very much inspired by Gauss's great work in that field, and Dirichlet's own book, the *Vorlesungen über Zahlentheorie*

(1863; Lectures on Number Theory), is of comparable historical importance to Gauss's *Disquisitiones.* He made many important discoveries in the field and his work on a problem connected with primes led him to make the fundamentally important innovation of using analytical techniques to obtain results in number theory. His stay in Paris had stimulated Dirichlet's interest in Fourier series and in 1829 he was able to solve the outstanding problem of stating the conditions sufficient for a Fourier series to converge. (The other problem of giving necessary conditions is still unsolved). Fourier also gave the young Dirichlet an interest in mathematical physics, which led him to important work on multiple integrals and the boundary-value problem, now known as the *Dirichlet problem,* concerning the formulation and solution of those partial differential equations occurring in the study of heat flow and electrostatics. These are of great importance in many other areas of physics. The growth of a more rigorous understanding of analysis owes to Dirichlet what is essentially the modern definition of the concept of a function. His *Vorlesungen über Zahlentheorie* contained important results about IDEALS and provided a lucid exposition of Gauss' results in number theory. He made important advances in the fields of number theory, complex analysis, mechanics, and the study of FOURIER SERIES, and introduced the modern concept of a FUNCTION a many-one relation.

Doodson, Arthur Thomas. (b. Mar. 31, 1890; Worsley, England; d. Jan. 10, 1968; Birkenhead, England). British mathematical physicist. Doodson, the son of a manager of a cotton mill, was educated at the University of Liverpool. After working at University College, London, from 1916 to 1918 he joined the Tidal Institute, Liverpool, in 1919 as its secretary. Doodson remained through its re-formation as the Liverpool Observatory and Tidal Institute as assistant director (1929-45) and as director until his retirement in 1960. Much of Doodson's early work was on the production of mathematical tables and the calculation of trajectories for artillery. Proving himself an ingenious, powerful, and practical mathematician he found an ideal subject for his talents in the complicated behavior of the tides. He made many innovations in their accurate computation and, with H. Warburg in 1942, produced the *Admiralty Manual of Tides.*

E

Eddington, Sir Arthur Stanley. (b. Dec. 28, 1882; Kendal, England; d. Nov. 22, 1944; Cambridge; England). British astrophysicist and mathematician. Eddington's father died in 1884 and he moved with his mother and sister to Somerset. He was a brilliant scholar, graduating from Owens College (now the University of Manchester) in 1902 and from Cambridge University in 1905. From 1906 to 1913 he was chief assistant to the Astronomer Royal at Greenwich after which he returned to Cambridge as Plumian Professor of Astronomy. He was knighted in 1930. Eddington was a Quaker throughout his life. Eddington was the major British astronomer of the interwar period. His early work on the motions of stars was followed, from 1916 onward, by his work on the interior of stars, which was published in his first major book, *The Internal Constitution of the Stars* (1926). He introduced "a phenomenon ignored in early investigations, which may have considerable effect on the equilibrium of a star, viz. the pressure of radiation." He showed that for equilibrium to be maintained in a star, the inwardly directed force of gravitation must be balanced by the outwardly directed forces of both gas pressure and radiation pressure. He also proposed that heat energy was transported from the center to the outer regions of a star not by convection, as thought hitherto, but by radiation. It was in this work that Eddington gave a full account of his mass-luminosity relationship, which was discovered in 1924 and shows that the more massive a star the more luminous it will be. The value of the relation is that it allows the mass of a star to be determined if its intrinsic brightness is known. This is of considerable significance since only the masses of binary stars can be directly calculated. Eddington realized that there was a limit to the size of stars: relatively few would have masses exceeding 10 times the mass of the Sun while any exceeding 50 solar masses would be unstable owing to excessive radiation pressure. Eddington wrote a number of books for both scientists and laymen. His more popular books, including *The Expanding Universe* (1933), were widely read, went through many editions, and opened new worlds to many enquiring minds of the interwar years. It was through Eddington that Einstein's general theory of relativity reached the English-speaking world. He was greatly impressed by the theory and was able to provide experimental evidence for it. He observed the total solar eclipse of 1919 and submitted a report that captured the intellectual imagination of his generation. He reported that a very precise and unexpected prediction made by Einstein in his general theory had been successfully observed; this was the very slight bending of light by the gravitational field of a star — the Sun. Further support came in 1924 when Einstein's prediction of the reddening of starlight by the gravitational field of the star was tested: at Eddington's request Walter Adams detected and measured the shift in wavelength of the spectral lines of Sirius B, the dense white-dwarf companion of the star Sirius. Eddington thus did much to establish Einstein's theory on a sound and rigorous foundation and gave a very fine presentation of the subject in his *Mathematical Theory of Relativity* (1923).Eddington also worked for many years on an obscure but challenging theory, which was only published in his posthumous work, *Fundamental Theory* (1946). Basically, he claimed that the fundamental constants of science, such as the mass of the proton and the mass and charge of the electron, were a "natural and complete specification for constructing a universe" and that their values were not accidental. He then set out to develop a theory from which such values would follow as a consequence, but never completed it.

Ehrenfest, Paul. (b. Vienna, Austria-Hungary [now Austria], 18.1.1880; d. Amsterdam, The Netherlands, 25.9.1933). Austrian theoretical physicist and mathematician who specialized mainly in statistical mechanics. Stimulated by working with BOLTZMANN, and by a period in Göttingen, he studied PLANCK's theory of black-body radiation and its relation to the foundations of statistical mechanics. He tackled the basic problem of reconciling the reversibility of classical mechanics with the irreversibility of the events of ordinary experience, showing that this is due to the extreme improbability of processes in which order spontaneously increases. He clarified many of the basic assumptions of statistical mechanics, and showed that it is logically necessary for the energy to have only certain discrete allowed values. In an analysis of the motion of a wave packet, he showed that its position and momentum obey the same equations of motion as a classical particle (Ehrenfest's theorem). His genius lay not in creation or in calculation but in criticism. He strove to understand the essence of a theory and to express it as clearly as possible. He knew what the problems were, he sharpened up the issues and asked the right questions. He was never satisfied with his work and always felt insecure and inadequate. This, together with the persecution of the Jews by the Nazis, led him to take his own life.

Eratosthenes. More than two thousand years ago Eratosthenes succeeded in measuring the circumference of the Earth. Eratosthenes was a Greek philosopher and man of learning and was especially recognised as a geographer and scientist. This man of genius was born in the year 276 B.C. He spent his youth studying under famous masters in different Greek centres of learning and especially in Athens. He became famous when still young and as a result he was asked to manage the library at Alexandria. This library was the most famous and the largest in the ancient world. He was also at this time appointed to be tutor to the Pharaohs, at the Egyptian court. He wrote about philosophy, drama, poetry and mathematics and geography, but unfortunately very few of these writings have survived. Eratosthenes drew the first complete geographic map of the inhabited world and as he was convinced of the spherical shape of the earth he suggested that it was possible to reach India by sailing westwards from Spain. He also thought that there were other inhabited lands opposite his own and in this way guessed at the existence of the American continent 1700 years before it was actually discovered. Even Christopher Columbus did not realise the existence of a new continent when he first set foot on it. At the age of 80 Eratosthenes was blind and weary of life. He therefore committed suicide by voluntary starvation. There follows an account of how he arrived at the most famous of his calculations: that of measuring the circumference of the earth. Alexandria and Siene are two cities in Egypt. Siene is near the present-day town of Aswan and is on the equator. Both Alexandria and Siene are on the same meridian. They are about 500 miles apart. (1) Siene, 21st June mid-day. The sun is perpendicular to the ground (shown by the fact that it is reflected in the water of wells) so its rays are directed towards the centre of the earth. (2) Alexandria is not on the equator, so the sun's rays are not vertical to the earth. (3) Knowing the height of an obelisk and by measuring the length of its shadows, one can calculate the inclination of the sun's rays, something of the order of 7°12'. (4) Now the ratio of 7°12' to one revolution of the sun (i.e. 360°) is going to be proportional to the distance between Alexandria and Siene and the distance (circumference) around the earth. 7°12' is a fiftieth of 360; therefore the circumference of the earth is 50x500 (the distance between Siene and Alexandria); this equals 25,000. The actual circumference of the earth is very close to this figure and depends which way round it is measured. Some authorities differ as to how correct Eratosthenes' actual calculations were. Their estimation lies between 2 and 20 per cent., depending on the value of the old measurement known as the stadium.

Ernest Nagel. Ernest Nagel was an American philosopher who was born in Nove Mesto, Czechoslovakia, in 1901. He came to the United States at the age of ten and was naturalized in 1919. He was educated in New York at City College and Columbia University, where he received his doctorate in philosophy in 1930. He taught at Columbia from 1931 until his retirement in 1961. Nagel collaborated in 1931 with Morris Cohen on his first book, *An Introduction to Logic and Scientific Method.* His later books dealt mainly with logic and the philosophy of science; he is perhaps best known for his book *The Structure of Science,* which offers his views on the logical structure of explanation. scientific inquiry, and the organization of scientific knowledge. Nagel was a Professor Emeritus at the time of his death in 1985. The recipient of many honors during his lifetime, he was highly regarded by his colleagues and exercised an extraordinary influence on the development of the philosophy of science in the twentieth century: "He has been a most influential teacher and an ideal intellectual companion to many philosophers, scientists, and men of affairs; he has not only exhibited but also rekindled in others commitments to clarity and concern for truth and for the need to integrate vision and technique." Nagel believed that philosophers could not formulate useful metaphysical ideas about the Universe unless they were familiar with the scientific definitions given to terms such as *space* and *time*. He regarded the need for logical consistency in the sciences not as an impediment which shackled the intellect but as a means for liberating the imagination because it enabled one to test the waters beforehand without indulging a reckless and essentially directionless philosophical inquiry. According to Nagel, "the conclusions of science are the fruits of an institutionalized system of inquiry." As a result, Nagel argued that logic and science are the necessary ingredients for a rational understanding of the Universe. Nagel's philosophical orientation aligned him with Einstein in his assertion that quantum mechanics is no less deterministic than classical mechanics. Nagel believed subatomic processes are not random events, arguing that the description of classical mechanics, which relies on identifying the "position" and "momentum" of particles, is not an appropriate tool for explaining quantum mechanics. He suggested that any series of events cannot be random because it corresponds to some mathematical function and therefore can be related to some set of laws. Nagel argued that confusion can result when a particular concept such as "particle" is transferred from one context, such as Newtonian mechanics, to another context, such as quantum theory, with no attention paid to the relationship of the concept itself to the context in which it is used.

Euelid. Geometry means literally 'measurement of the earth' — what we should now call 'surveying'. It arose out of the practical needs of the agricultural peoples of ancient Mesopotamia (modern Iraq) and Egypt, for they had to measure out and apportion their plots of land. In Egypt, the yearly flooding of the Nile, which fertilised the land as it still does today, meant that the boundaries of the fields had to be worked out again, after being washed away. These primitive ways of measuring spaces came next to be applied to building. The ancient temples of Egypt and Mesopotamia, their roads and canals, all bear witness to the geometrical skills their makers had attained. Records survive which show how far they had gone. Nevertheless, none of this is really 'geometry' as we know it. The word was coined by the Greeks, for it is they who first turned all the practical rules of measurement into an ordered branch of mathematics. To many generations of men Greek mathematics came to mean Euclid. Yet for nearly three centuries before his time there had been Greek mathematicians. The earliest mathematical proofs are attributed to Thales of Miletus, who lived about 585 B.C., whereas we know that Euclid was at work in 300 B.C. He taught at Alexandria, which lies at the western end of the Nile delta. The city had been enlarged and given its new name by Alexander the Great in 332 B.C., and was fast becoming the greatest centre of Greek influence outside Greece: what we now call 'Hellenistic culture'. It seems that Euclid was not so much an inventor as a recorder of the findings of others: as such he was the greatest mathematical schoolmaster of all time. His *Elements* is the only textbook that has ever been used for well over 2,000 years. Not until the 19th century did mathematicians show more precision than Euclid. Until then the *Elements* had been regarded as the most perfect example of a theory in which conclusions are drawn from assumptions in a systematic way: what we call a 'deductive theory'. From a small set of basic assumptions or 'axioms', with careful definitions and logical reasoning, he derived a vast array of theorems set out in an orderly way. To Euclid, the axioms were obvious: in no need of explanation. Today, we no longer feel bound to be as sure as he was. For instance, he took it that if two straight lines cutting a third make interior angles (angles on the same side) amounting to less than two right angles, then those two lines will eventually meet. This means that through any point outside a line there is one, and only one, line that will not meet the first: what is called a line 'parallel' to it. This for Euclid was an axiom. We no longer hold this to be obvious. For 200 years now mathematicians have tried to derive the parallel axiom from the other assumptions of Euclidean geometry, but in vain. If we simply drop the parallel axiom, we can build up other systems of geometrical theorems which hang together just as well. Such systems are called 'non-Euclidean', and they have become very important in modern science. Einstein's general theory of relativity makes use of one of these non-Euclidean geometries. As far back as the 6th century B.C. Pythagoras had discovered that the diagonal of a square could not be measured as an ordinary proportion of the side of the square. He called the diagonal 'unmeasurable', which the Romans translated into 'irrational', the word we use today. These irrational numbers were a great blow to the mathematicians of that time, because they had wrongly thought that all lengths could be measured as proportions that could be stated in whole numbers. However, they worked out a method of using whole numbers to get closer and closer to these measurements. But on the whole their methods for finding proportions were geometrical Euclid's 'Elements' deal with them at length. They also deal with arithmetic, and provide proofs for some of the basic elementary theorems about numbers. Especially famous is a proof about prime numbers (those numbers that have no factors): Euclid showed that, however big a prime number might be, there are always still bigger ones. Euclid also wrote on other subjects besides mathematics. Many of his books are lost, but his texts on optics and on music have survived. Euclid is one of the best known and most influential of classical Greek mathematicians but almost nothing is known about his life. He was a founder and member of the academy in Alexandria, and may have been a pupil of Plato in Athens. Despite his great fame Euclid was not one of the greates of Greek mathematicians and not of the same caliber as Archimedes. Euclid's most celebrated work is the *Elements*, which is primarily a treatise on geometry contained in 13 books. The influence of this work not only on the future development of geometry, mathematics, and science, but on the whole of Western thought is hard to exaggerate. Some idea of the importance that has been attached to the *Elements* is gained from the fact that there have probably been more commentaries written on it than on the Bible. The *Elements* systematized and organized the work of many previous Greek geometers, such as Theaetetus and Eudoxus, as well as containing many new discoveries that Euclid had made himself. Although mainly concerned with geometry it also deals with such topics as number theory and the theory of irrational quantities. One of the most celebrated number theoretic results is Euclid's proof that there is an infinite number of primes. The *Elements* is in many ways a synthesis and culmination of Greek mathematics. Euclid and Apollonius of Perga were the last Greek mathematicians of any distinction, and after their time Greek civilization as a whole soon became decadent and sterile. Euclid's *Elements* owed its enormously high status to a number of reasons. The most influential single feature was Euclid's use of the axiomatic method whereby all the theorems were laid out as deductions from certain self-evident basic propositions or axioms in such a way that in each successive proof only propositions already proved or axioms were used. This became accepted as the paradigmatically rigorous way of setting out any body of knowledge, and attempts were made to apply it not just to mathematics, but to natural science, theology, and even philosophy and ethics. However, despite being revered as an almost perfect example of rigorous thinking for almost 2000 years there are considerable defects in Euclid's reasoning. A number of his proofs were found to contain mistakes, the status of the initial axioms themselves was increasingly considered to be problematic, and the definitions of much basic terms as 'line' and 'point' were found to be unsatisfactory. The most celebrated case is that of the parallel axiom, which states that there is only one straight line passing through a given point and parallel to a given straight line. The status of this axiom was long recognized as problematic, and many unsuccessful attempts were made to deduce it from the remaining axioms. The question was only settled in the 19th century when Janos Bolyai and Nicolai Lobachevski showed that it was perfectly possible to construct a consistent geometry in which Euclid's other axioms were true but in which the parallel axiom was false. This epoch-making discovery displaced Euclidean geometry from the privileged position it had occupied. The question of the relation of Euclid's geometry to the properties of physical space had to wait until the early 20th century for a full answer. Until then it was believed that Euclid's geometry gave a fully accurate description of physical space. No less a thinker than Immanuel Kant had thought that it was logically impossible for space to obey any other geometry. However when Albert Einstein developed his theory of relativity he found that the appropriate geometry for space was not Euclid's but that developed by Georg Riemann. It was subsequently experimentally verified that the geometry of space is indeed non-Euclidean. In mathematical terms too, the discovery of non-Euclidean geometries was of great importance, since it led to a broadening of the conception of geometry and the development by such mathematicians as Felix Klein of many new geometries very different from Euclid's. It also made mathematicians scrutinize the logical structure of Euclid's geometry far more closely and in 1899 David Hilbert at last gave a definitively rigorous axiomatic treatment of geometry and made an exhaustive investigation of the relations of dependence and independence between the axioms, and of the consistency of the various possible geometries so produced. Euclid wrote a number of other works besides the *Elements,* although many of them are now lost and known only through references to them by other classical authors. Those that do survive include *Data,* containing 94 propositions, *On Divisions,* and the *Optics*. One of his sayings has come down to us. When asked by Ptolemy I Soter, the reigning king of Egypt, if there was any quicker way to master geometry than by studying the *Elements* Euclid replied "There is no royal road to geometry."

Eudoxus of Cnidus. (b. c. 400 B.C.; Cnidus, now in Turkey; d. c. 350 B.C.; Cnidus). Greek astronomer and mathematician. Eudoxus is reported as having studied mathematics under Archytas, a Pythagorean. He also studied under Plato and in Egypt. Although none of his works have survived they are quoted extensively by Hipparchus. Eudoxus was the first astronomer who had a complete understanding of the celestial sphere. It is only this understanding that reveals the irregularities of the movements of the planets that must be taken into account in giving an accurate description of the heavens. For Eudoxus the Earth was at rest and around this center 27 concentric spheres rotated. The outermost sphere carried the fixed stars, each of the planets required four spheres, and the Sun and the Moon three each. All these spheres were necessary to account for the daily and annual relative motions of the heavenly bodies. He also described the constellations and the changes in the rising and setting of the fixed stars in the course of a year. In mathematics, Eudoxus is thought to have contributed the theory of proportion to be found in Book V of Euclid — the importance of this being its applicability to irrational as well as rational numbers. The method of exhaustion in Book XII is also attributed to Eudoxus. This tackled in a mathematical way for the first time the difficult problem of calculating an area bounded by a curve.

Eugene P. Wigner. When Eugene Wingner received the Nobel Prize in physics in 1965, the citation mentioned no particular scientific accomplishment to justify the award. Instead it referred vaguely to the many areas in modern physics in which Wigner had made important contributions. Among other things Wigner is a fine writer of nontechnical essays, many of which are collected in his book *Symmetries* and *Reflections*. His language is sparse and precise and his essays often reflect a philosophical bent whether he is discussing the physics of life or the effectiveness of mathematics. Mathematics, for Wigner, is an extremely useful tool for analyzing and describing physical phenomena. But he says he is unable to explain why mathematics is so useful to the natural sciences. However, he sees mathematics as perhaps the supreme example of human ingenuity because "the mathematician could formulate only a handful of interesting theorems without defining concepts beyond those contained in the axioms and that the concepts outside those contained in the axioms are defined with a view of permitting ingenious logical operations which appeal to our aesthetic sense both as operations and also in the results of great generality and simplicity." Wigner regards the laws of nature as invariant since they do not change as the position of the experimenter changes. In other words, a stone dropped from a tower in Europe will fall no faster or slower than a stone dropped from a similar-sized tower anywhere else in the world. More important, laws of nature are constant throughout the Universe. The thermonuclear reactions which power the stars are physically invariant throughout all the galaxies of the Universe. This regularity or predictability is a common element in both physics and mathematics. These laws of nature are a succession of layers with "each layer containing more general and more encompassing laws than the previous one and its discovery constituting a deeper penetration into the structure of the Universe than the layers recognized before." As a result, Wigner's view of physics in general seems to be that it is a cumulative process, most of which will ultimatley be consigned to the theoretical scrapheap as better theories are uncovered. Laws of nature are conditional statements and their applicability will always be limited to a very small part of the physical world. However, mathematics is the form in which these laws are described and as such is essential to physics. With the aid of mathematics, a surprisingly wide array of phenomena can be deduced, described, and explained from basic laws, only a small portion of which might have been originally surmised without mathematics.

Euler, Leonhard. (b. Apr. 15, 1707; Basel, Switzerland; d. Sept. 18, 1783; St. Petersburg, now Leningrad in the Soviet Union). Swiss mathematician. Euler was one of the outstanding figures of 18th century mathematics, and also the most prolific. He studied at the University of Basel where he came to the notice of the Bernoulli family and became, in particular, a friend of Daniel Bernoulli. Having completed his studies Euler applied unsuccessfully for a post at Basel University. He was encouraged by his friend Bernoulli to join him at the St. Petersburg Academy of Science. Bernoulli obtained for Euler a post in the medical section of the academy and in 1727 Euler left for Russia to find on arriving that the empress had just died and that the future of the academy was doubtful. Fortunately it survived intact and he managed to find his way into the mathematical section. In 1733 Bernoulli left Russia to return to Switzerland and Euler was appointed to replace him as head of the mathematical section of the academy. Euler was not particularly happy in the highly repressive political climate and devoted himself almost exclusively to his mathematics. In 1740 he eventually left Russia to join Frederick the Great's Berlin Academy. He returned to Russia in 1766 at the invitation of Catherine the Great, remaining there for the rest of his life. He had lost the sight of his right eye through observing the Sun during his first stay in Russia and shortly after returning there his blindness became total. However, such was his facility for mental calculation and the power of his memory that this did not affect his mathematical creativity. Euler contributed to almost all areas of mathematics, and did equally important work in both pure and applied mathematics. One of his most significant contributions was to the development of analysis. Although he was working before the development of modern standards of rigor by such mathematicians as Karl Friedrich Gauss and Augustin Cauchy and thus lacked a rigorcus treatment of such key topics as convergence, nonetheless his work in analysis constitutes a major advance over previous work in the area. He wrote three treatises on different aspects of analysis, which together collect, systematize, and develop what the mathematicians of the 17th century had achieved. These treatises became important and influential textbooks. It is worth noting that unlike some great mathematicians, such as Gauss, Euler was a highly successful and effective teacher. it is a measure of his intuitive insight into mathematics that even though he did lack truly rigorous analytical techniques he was still able to arrive at so many novel and important results. His phenomenal ability for calculation came to his aid here, for he frequently arrived at results, for example about infinite series, by induction from a great many calculations and gave only a highly dubious proof leaving future mathematicians to give properly rigorous proofs of results that were indeed quite correct. Outside analysis Euler made extremely important contributions both to the calculus of variations and mechanics. Mechanics was transformed by his treatment of it in his treatise of 1736. In essence he transformed the subject into one to which the full resources of analytical techniques could be applied. In doing so he paved the way for the work of mathematicians such as Joseph Lagrange. Euler put his expertise in mechanics to practical use in his work on the three-body problem. He was interested in this problem because he wished to investigate the motion of the Moon. He published his first lunar-motion theory in 1753 and then a second theory in 1772. Euler was able to invent methods of approximating solutions that were accurate enough to be of practical use in navigation. Here his prodigious ability as a sheer calculator came into its own. In addition to these pieces of work Euler made notable contributions to number theory and geometry. Numerous theorems and methods are named for him. He was renowned for his ability to perform complex calculations mentally, and continued his work even after he lost his sight. He was one of the most prolific mathematicians of all time, publishing over 500 papers and textbooks on virtually all branches of mathematics (a further 350 have appeared posthumously). His most significant contributions were to analytical geometry, calculus and trigonometry, and thereby to the unification and systematization of all of mathematics.

F

Fermat, Pierre de. (b. Aug. 17, 1601; Lomagne, France; d. Jan. 12, 1665; Castres, France). French mathematician and physicist. Fermat was one of the leading mathematicians of the early 17th century although not a professional mathematician. He studied law and spent his working life as a magistrate in the small provincial town of Castres. Although mathematics was only a spare-time activity. Fermat was an extremely creative and original mathematician who opened up whole new fields of enquiry. Fermat's work in algebra built on and greatly developed the then new theory of equations, which had been largely founded by Francois Viète. With Pascal, Fermat stands as one of the founders of the mathematical theory of probability. In his work on methods of finding tangents to curves and their maxima and minima he anticipated some of the central concepts of Isaac Newton's and Gottfried Leibniz's differential calculus. Another area of mathematics that Fermat played a major role in founding, independently of Rene Descartes, was analytical geometry. This work led to violent controversies over questions of priority with Descartes. Nor were Fermat's disagreements with Descartes limited to mathematics. Descartes had produced a major treatise on optics — the *Dioptrics* — which Fermat greatly disliked. He particularly objected to

Descartes' attempt to reach conclusions about the physical sciences by purely a *priori* rationalistic reasoning without due regard for empirical observation. By contrast Fermat's view of science was grounded in a thoroughly empirical and observational approach, and to demonstrate the errors of Descartes' ways he set about experimental work in optics himself. Among the important contributions that Fermat made to optics are his discovery that light travels more slowly in a denser medium, and his formulation of the principle that light always takes the quickest path. Fermat is probably best known for his work in number theory, and he made numerous important discoveries in this field. But he also left one of the famous unsolved problems of mathematics — *Fermat's last theorem*. This theorem states that the algebraic analog of Pythagoras's theorem has no whole number solution for a power greater than 2, i.e. the equation $a^n + b^n = c^n$ has no solutions for n greater than 2, if *a, b,* and *c* are all integers. Fermat himself thought he had found a proof of this result, but this proof was subsequently lost. As mathematicians' attempts to re-prove the 'last theorem' since then have been unsuccessful, it is generally thought that Fermat's proof must have been mistaken. He is credited with founding modern number theory and (independently of Pascal) the calculus of probability, as well as discovering analytical geometry independently of Descartes. He obtained sophisticated results in the foundations of analytic geometry and differential calculus, but failed to publish them. He claimed to have proved the famous unsolved problem known as Fermat's Last Theorem.

Fibonacci, Leonardo. (b. *c.* 1170; d. *c.* 1250). Italian mathematician. Fibonacci lived in Pisa and is often referred to as Leonardo of Pisa. Although he was probably the most outstanding mathematician of the Middle Ages virtually nothing is known of his life. The modern system of numerals, which originated in India and had first been introduced to the West by al-Khwarizmi, first became widely used in Europe owing to Fibonacci's popularization of it. His father served as a consul in North Africa and it is known that Fibonacci studied with an Arabian mathematician in his youth, from whom he probably learned the decimal system of notation. Fibonacci's main work was his *Liber abaci* (1202; Book of the Abacus) in which he expounded the virtues of the new system of numerals and showed how they could be used to simplify highly complex calculations. Fibonacci also worked extensively on the theory of proportion and on techniques for determining the roots of equations, and included a treatment of these subjects in the *Liber abaci*. In addition it contains contributions to geometry and Fibonacci later published his *Practica geometrine* (1220; Practice of Geometry), a shorter work that was devoted entirely to the subject. Fibonacci was fortunate in being able to gain the patronage of the Holy Roman Emperor, Frederick II, and a later work, the *Liber quandratorum* (1225; Book of Square Numbers) was dedicated to his patron. This book, which is generally considered Fibonacci's greatest achievement, deals with second order Diophantine equations. It contains the most advanced contributions to number theory since the work of Diophantus, which were not to be equalled until the work of Fermat. He discovered the 'Fibonacci sequence' of integers in which each number is equal to the sum of the preceding two (1, 1, 2, 3, 5, 8,...).

Fisher, Ronald Aylmer. (b. London, UK, 17.2.1890; d.

Adelaide, S. Australia, Australia, 29.7.1962). British statistician and geneticist. Fisher worked at the Agricultural Research Station at Rothamstead (1919-33) before becoming professor of eugenics in London, and, later, of genetics at Cambridge. He first came to notice by his derivation of the sampling distribution of the correlation coefficient given independence. He later added distributions for the non-independent case, for partial and multiple correlations, and for means of small samples, mean deviations, variance ratios and higher-order moments. His work on estimation and inference was based on the likelihood function, expressing the variation, over the possible values of the underlying parameters, of the probability of the data actually observed. His principle of maximum likelihood estimation was to choose those values that maximized that probability. He developed analytical means to this end, especially the concept of a sufficient statistic for a parameter, which is a suitable quantity summarizing all the information in a sample that has any bearing on that parameter. If data, coming from several sources, are all drawn from one population, the variation of the group means will reflect the population variance, and so will the variation within the groups. If not, there will be a discrepancy which can be assessed by a variance ratio test. This technique of the analysis of variance was made ubiquitous by Fisher, with simultaneous assessments of many potential sources of variation being made by way of orthogonal components. In studying responses of, say, fruit trees to fertilizers, few of those variables that might obscure the truth can be controlled, and those left are too numerous to be accounted for numerically. Fisher realized that if each experiment has the layout of the trees, together with the assignment of fertilizers, chosen at random from among all viable designs, then the chance of a persistent fortuitous result can be virtually eliminated. This principle of randomization (sometimes misinterpreted as an imperative that can override ethics) started the modern science of experimental design (see his book *The Design of Experiments,* London, 1938). Fisher had always been interested in biology and he made significant contributions towards the solving of two central problems in genetics: the hereditary determination of continuous variation, and the relationship between Darwin's theory of natural selection and MENDEL's principles of heredity. Many biometricians had maintained that the theory of blending inheritance was more applicable to continuous variation than was the theory of particulate inheritance. But the possibility that continuous variation could be the result of many genes affecting the same character had been advanced by a number of people (including W. Weinberg, who anticipated Fisher in his approach). By analysing correlations between human relatives, Fisher clearly showed not only that these correlations could be interpreted according to Mendelian inheritance, but that Mendelian inheritance must lead to the observed correlations (see 'The correlation between relatives on the supposition of Mendelian inheritance', 1918). The theory of blending inheritance had been advocated earlier by Darwin, but led to difficulties in that it predicts a reduction in the genetic variation on which natural selection acts. That Mendelian heredity provides an answer to the problem of the origin of genetic variation was not fully appreciated until Fisher pointed out that it conserves this, and furthermore that mutations do not themselves direct the course of evolution (see *The Genetical Theory of Natural Selection,* Oxford, 1930).

The development by Fisher of his fundamental theorem of natural selection is one of the cornerstones of population and evolutionary genetics.

Fourier, Jean Baptiste Joseph, Baron de (1768-1830). French analyst and physicist, whose study of the conduction of heat had a profound influence on mathematical physics and on the study of real functions. Of humble origin, he became a professor in a military academy, accompanied Napoleon on his Egyptian expedition and was appointed Governor of Lower Egypt; after the French defeat there, he became Prefect of Grénoble and was created a baron. He also published widely on Egyptology, became permanent secretary of the French Academy of Science, and was elected to the Academy of Medicine, the *Académie Francaise,* and the Royal Society. Fourier, the son of a tailor, was educated at the local military school and later at the Ecole Normale in Paris. He held posts at both the Ecole Normale and the Ecole Polytechnique where he was a very effective and influential teacher. In 1798 he accompanied Napoleon on the invasion of Egypt and later contributed to and oversaw the publication of the *Description de l'Egypte* (1808-25), a massive compilation of the cultural and scientific materials brought back from the expedition. Fourier's most important mathematical work is contained in his *Théorie analytique de la chaleur* (1822; The Analytical Theory of Heat), a pioneering analysis of the conduction of heat in solid bodies in terms of infinite trigonometric series, now known as *Fourier series.* Fourier was led to consider these series when attempting to solve certain boundary-value problems in physics and his interest was always in the physical applications of mathematics rather than in its development for its own sake. His work continues to be extremely important in many areas of mathematical physics, but it has also been developed and generalized to yield a whole new branch of mathematical analysis, namely, the theory of harmonic analysis.

Fréchet, Maurice-René. (b. Maligny, Yonne, France, 10.9.1878; d. Paris, 4.6.1973). French mathematician who helped found the theory of abstract spaces and set-theoretical or general topology and contributed to the development of functional and abstract analysis. When studying mathematics at the University of Paris he was encouraged to examine the emerging field of functional analysis, a generalization of classical analysis, by his thesis supervisor HADAMARD. The result was his important and influential doctoral thesis 'Sur quelque points du Calcul fonctionnel' [Some aspects of functional calculus] of 1906, in which he initiated the study of various abstract spaces from a set-theoretical viewpoint. The most important type of space was the (E) class or metric space as Hausdorff later called it. He also isolated the concept of compactness as an abstract notion and investigated various function spaces and functionals. He then applied the results to branches of analysis. Around 1909-10 he created an interesting topological theory of dimension. His early work on abstract spaces influenced many mathematicians who were developing analysis from an abstract viewpoint. His work was generalized by F. Hausdorff in the general theory of topological spaces given in Hausdorff's *Fundamentals of Quantity Theory* (1914). Up to 1928 Frechet continued to investigate many types of abstract spaces and many parts of abstract analysis; he summarized his work in his most important book *Les Espaces abstraits* [Abstract spaces] (Paris, 1928). After that time he turned to the study of probability theory. The kind of abstract mathematics which Fréchet helped to develop has been very much in vogue during the 20c.

Fredholm, Erik Ivar. (b. Apr. 7, 1866; Stockholm; d. Aug. 17, 1927; Stockholm). Swedish mathematician. Fredholm, who was a pioneer in the theory of integral equations, studied at the Stockholm Polytechnic Institute, the University of Stockholm and the University of Uppsala, from which he received his doctorate in 1898. The same year he was appointed lecturer in mathematical physics at the University of Stockholm, becoming professor there in 1906. In the field of mathematical physics Fredholm's work on the deformation of anisotropic media (such as crystals) led him to the study of partial differential equations. His most important work was on integral equations establishing the field in which David Hilbert was to do some of his greatest work. A number of results and concepts are named for him including the *Fredholm integral equations* and the *Fredholm operator.*

Frege, Gottlob. (b. Wismar, Germany [now East Germany], 8.11.1848; d. Bad Kleinen, 26.7.1925). German philosopher and mathematician. The son of a protestant clergyman, Frege spent his whole professional career at the University of Jena. He carried on his work in isolation and comparative obscurity, although in the first decade of the century his importance was recognized by Husserl, who creditably acknowledged the authority of a fierce review of his first book (*Philosophie der Arithmetik,* 1891) by Frege, and by Bertrand Russell, who devoted an appendix in *The Principles of Mathematics* (1903) to Frege's ideas, having arrived himself at a position much like Frege's, although less rigorously thought out. And some 40 years after the end of his main productive period, Frege was mentioned admiringly in Wittgenstein's *Tractatus* (1922). At a time when pure mathematics was becoming ever more abstract and remote from intuition. Frege saw the need for a thorough investigation, to be conducted with the greatest possible degree of logical rigour, both of the foundations of the discipline and of its subsequent development. The first step in this project was his working out of a new formal notation for logic in *Begriffschrift* (Halle, 1879; tr. T.W. Bynum in *Conceptual Notation and Related Articles,* Oxford, 1972). But Frege's choice of a complex idiographic notation contributed to his consequent neglect by its lack of perspicuity. In his *Die Grundlagen der Arithmetik* (Breslau, 1884; tr. J.L. Austin, *The Foundations of Arithmetic,* Oxford, 1950, [2]1953) he argued for the 'logistic' thesis (independently arrived at by Russell some 20 years later) that mathematics is really a direct continuation of, or deduction from, an enriched formal logic. Crucial to the thesis is his definition of number in terms of classes or sets. The importance of this definition was undermined by Russell's discovery, communicated to Frege in a letter in June 1902, of the paradox that the class of classes that are not members of themselves both is and is not a member of itself. Frege replied, without irony, 'arithmetic totters'. He had already worked out the derivation via classes or sets in detail in his *Die Grundgesetze der Arithmetik* (2 vols, Jena, 1893-1903; tr. M. Furth, *The Basic Laws of Arithmetic,* Berkeley, Calif., 1964). The logic from which Frege began the derivation had as a vital constituent predicate logic or quantification theory, an incomparably more powerful and comprehensive account of the relations between statements about all, some or none of the things of a given kind than is to be found in Aristotle's theory of the syllogism which plays only a small, elementary part in Frege's system. Frege went on, in a series of essays still profoundly influential in their own right and not simply as the initiators of later developments, to transform the philosophical understanding of the basic notions of logic, most importantly in 'Ober Sinn und Bedeutung' [On sense and reference] (1892; tr. P. Geech & M. Black in *Translations from the Philosophical Writings of Gottlob Frege,* Oxford, ²1960) in which the distinction between meaning proper and reference, often previously recognized in principle, was examined with unprecedented thoroughness and penetration. Here, as in his other writings, Frege was a tireless and lethal critic of 'psychologism', the point of view which takes human thought processes to be the subject matter of logic and mathematics. He maintained that these exact disciplines concerned a third realm, distinct from both mind and physical nature, composed of timeless Platonic essences. Logicians today recognize Frege as the only exponent of their science to rank with Aristotle. His work remains unmatched for its combination of constructive fertility, rigour of execution and breadth of scope. He taught all branches of mathematics at Jena throughout his life, but almost all his publications were logical. His main innovations were the distinction between Sense and Rerference, and a logic of quantification, in which the Quantifiers are treated as properties of properties and for which he devised an idiosyncratic 'concept notation' (Begriffsschrift) so elaborate as to discourage understanding; although many of his ideas became familiar through the work of Peano and Russell, it was their notation which became standard. He wrote on the foundations of mathematics, setting out axioms for set theory from which he believed arithmetic could be derived. His bitterness about the indifference and hostility with which his work was received was increased by a sarcastic review by Cantor, who had not even troubled to read the book; then, when the second volume of his development of Logicism was already in the press, he received a letter from Bertrand Russell, one of his few admirers, informing him that his axioms were inconsistent. Although Frege did revise his axioms (unsuccessfully), he abandoned the projected third volume, and was too depressed to do any other useful work; his bitterness was compounded by the fate of Germany in the First World War, and his diary reveals, as well as virulent antisemitism, a pathological hatred of Catholics, the French, socialism, and democracy. He did, however, publish three more philosophical papers; in his last years he became convinced of the falsehood of his logicism, but his revised views were never published. Frege is now regarded as a crucial figure in the history of both logic and philosophy, and he and Wittgenstein, whom he greatly influenced, are the source of almost all modern philosophy of language.

Frege, (Friedrich Ludwig) Gottlob. (b. Nov. 8, 1848; Wismar, now in East Germany; d. July 26, 1925; Bad Keinen, now in East Germany). German philosopher and mathematician. Frege studied at the universities of Jena and Göttingen, where he obtained his PhD in 1873. He then returned to Jena as a lecturer, where he remained for the rest of his working life, rising to the position of professor in 1896. In a series of seminal works Frege laid the foundations of modern mathematical logic, transforming logic with an understanding of and notation for the problem of multiple generality — propositions containing predicates, quantifiers, and variables — and showing how the basic concepts and operations of mathematics could be formalized. He also revolutionized modern philosophy through his influence on the philosophy of language. However Frege's work was almost completely ignored, misunderstood, or treated with hostility by his contemporaries — notable exceptions were Bertrand Russell and Giuseppe Peano. In *Die Grundlagen der Arithmetik* (1884; The Foundations of Arithmetic) Frege gave a formal definition of cardinal number and showed how basic properties of numbers could be logically derived from it. In *Grundgesetze der Arithmetik* (1893 and 1903; Basic Laws of Arithmetic) he went further in attempting to derive arithmetic from formal logic. The *Grundgesetze* is still regarded as a massive achievement, but his main aim was doomed to failure. On the eve of the publication of the second volume Russell wrote to Frege pointing out a contradiction — Russell's paradox (see Bertrand Russell) — that could be derived from his sytem. This, as Frege acknowledged, vitiated his whole project.

G

Galois, Evariste. (1811-32), French mathematician who made significant contributions to the theory of functions, the theory of equations, and number theory, and whose work became a basis for group theory (a term which he introduced); all this developed from his concern, while still at school, to show the impossibility of the solution by radicals of the quintic (which had, unknown to Galois, already been shown by Abel), and to describe the general conditions for the solubility of any polynomial equation. Although he had already published some papers, when he submitted work to the Academy of Sciences in 1829, the first was lost by Cauchy, and the second by Fourier; he also clashed with the oral examiner for the *École Polytechnique* and was refused a place. After his father committed suicide, he abandoned thoughts of mathematics as a career, and enrolled as a trainee teacher, only to be expelled for writing an anti-monarchist article, and he was imprisoned twice because of his republican belief. His third submission to the Academy was rejected by Poisson. Galois was killed in a duel, probably provoked by royalist or police agents, the age of 20, and is generally viewed as one of the two great romantic figures in mathematics) the other being Ramanujan). Galois was born during the rule of Napoleon. He entered the College Royale de Louis-le-Grand in Paris in 1823 and it was here that his precocious mathematical genius first emerged. He published several papers while still a student and at the age of about 16 embarked upon his noted work on algebraic equations. But his career was marred by lack of advancement, associated with political bitterness. Twice, in 1827 and 1829, he was rejected by the Paris Ecole Polytechnique, and three papers submitted to the Academy of Sciences were rejected or lost. In 1830 he entered the Ecole Normale Supérieure to train as a teacher. That year revolution in Paris caused the abdication of Charles X, who was succeeded by Louis Philippe. Galois — fiercely republican — was expelled for writing an antiroyalist newspaper letter. In 1831 he was arrested twice: once for a speech against the king and the second time for wearing an illegal uniform and carrying arms — for this he received six months imprisonment. In the spring of 1832 he died in a duel; the details are uncertain but it may have been provoked by political opponents. Galois seems to have anticipated that he was to die, for the night before was spent desperately recording his mathematical ideas in a letter to his former schoolmaster, Auguste Chevalier. Here he outlined his work on elliptic integrals and set out a theory of the roots (solutions) of equations, in which he considered the properties of permutations of the roots. Admissible permutations — ones in which the roots obey the same relations after permutation — form what is now as a *Galois group,* having properties that throw light on the solvability of the equations. The manuscripts were published in 1846 and his work recognized. With the equally tragic Norwegian, Niels Henrik Abel, he is regarded as the founder of modern group theory.

Galileo. Galileo Galilei was born at Pisa on February 18, 1564. He belonged to a noble family of Florence, whose original surname was Bonajuti, but who took the name of Galilei in 1343. Members of this family had filled many positions of honor in the state, but Galileo's father, Vincenzo de Bonajuti de Galilei, though a clutivated man, a musician and mathematician, was very poor. He desired, therefore, that his son's studies should be, unlike his own, in branches of learning which might prove materially remunerative, and Galileo, at seventeen years of age, was sent to the University of Pisa to study medicine. Up to this time he had been at school at the monastery of Vallombrosa, and had shown himself not only to be a quick and intelligent scholar, but also to have considerable taste and aptitude for poetry, music and painting. His father, knowing by experience that mathematics were a peculiarly unremunerative form of study, endeavored to prevent Galileo from becoming versed in them, for already, as a boy, his skill in devising mechanical toys showed that he had inherited considerable ability in this direction. However, early in 1583. Galileo, straying into a lecture on Euclid, became fascinated with the subject and began to devote all his spare time to it. His own disinclination to pursue the study of medicine and the difficulty which his father found in paying for his college education, added to the fact that his independnt and argumentative disposition made him unpopular with the authorities, resulted in his return, in 1589, to his father's house at Florence. Here Galileo pursued the study of mathematics with ardor and, in 1585, after several vain attempts to secure a similar position, was appointed lecturer in mathematics at the University of Pisa. His salary was absurd—sixty scudi a year, or about $ 65 of our money. He was obliged to leave Pisa on account of the storm of hatred aroused by his discovery and teaching of the laws of falling bodies. Aristotle had taught that bodies fall in a time proportional to their weight, but Galileo showed, by dropping various bodies from the Leaning Tower of Pisa, how false this assumption was, and stated the real laws of falling bodies. This discovery is the most important of all Galileo's contributions to science, for upon the laws of motion which he expounded the whole science of kinetics, as now understood, virtually rests. But it lost him favor and position, so unwilling were the mathematicians to be convinced by demonstration against accepted authority. In 1592 Galileo was appointed professor of mathematics at Padua University for six years, and this appointment was twice renewed, so that he retained the position until 1610. In 1597 he invented his proportional compasses, the earliest representative of the sextant. In 1604 the appearance of a brilliant new star in Serpentarius gave Galileo the opportunity of launching his shafts against the accepted astronomical theories. He attacked first of all the Aristotelian teaching, universally held at that time, as to the unchanging and incorruptible character of the heavens; and proceeded, with all the fervor and probably much of the bold assumption of the new convert, to set forth the Copernican system of the heliocentric universe, and to repudiate the Ptolemaic or geocentric system. In 1609, hearing of the invention of a Dutch optician for bringing distant objects near, Galileo in one night succeeded in working out the principle, and at once set to work to construct a telescope and, having produced a satisfactory one, turned it towards the heavens, with results which in his day were truly astonishing. For the first time sun-spots came to sight, supplying Galileo with a powerful argument against the doctrine of immutability, and by their movement across the face of the sun indicating its rotatory motion. With it, too, Galileo discovered Jupiter's satellites, and proved the truth of Copernicus's assertion that, if we could but see them clearly. Venus and Mercury would be found to show phases similar to those of the moon. Galileo was triumphant. Overwhelming proof seemed to him to be now provided both against the Ptolemaists and against the supporters of the theory of immutability. The telescope was Galileo's most famous gift to science, though it really was not so important as his discovery of the laws of motion; for the telescope would inevitably have been improved and turned upon the heavens — which was Galileo's share in its discovery — before long, even without him, whereas it is unlikely that the laws of motion would have been truly expounded for some time to come. It is frequently supposed, by those unfamiliar with the more recent historical investigations on this subject, that the Church had set her face utterly against scientific progress, and in particular against Galileo's discoveries. It is certain, however, that, in the year 1611 Galileo visited Rome, and was received with great honor by Pope, cardinals and priests. His telescope was set up in the Quirinal garden, the property of Cardinal Bandini, and crowds of curious and interested spectators were allowds to come there and look through it. Trouble certainly arose, however, when Galileo, always assertive and dogmatic, rendered arrogant by his fame and success, went beyond his own sphere and maintained that Scripture was on his side. In order to realize how dangerous such an assertion appeared to ecclesiastics to be to the faith of the peopole, we need to remember the familiar association between the Scriptural accounts and teaching and the accepted Ptolemaic system, which, however scientifically false, still remains descriptively true as a mere explanation of appearances. It is said in defense of the Church that it was ready to listen so long as Galileo condfined himself to teaching the Copernican system merely as a hypothetical explanation, which it was for future science carefully to determine by observation and inquiry, but that when Galileo asserted it as absolute truth and as the real teaching of Scripture, they repudiated and condemned his teaching. The proofs which were at that time adduced in support of the Copernican system were by no means eminently satisfying, and many of them are now known to be false. On such inadequate grounds a theory of the universe was boldly proclaimed to the people which was almost certain to result in undermining their faith, since to the ignorant mind it impugned the authority of Scripture. The Church was not concerned with questions of science still manifestly theoretical; she was deeply concerned in the faith of the people. Hence Galileo got into trouble with the Church. We are told that he was accused berfore the Inquisition in 1615, but at that time the result was that he was merely recommended to confine himself to his mathematical reasonings upon this system, and to abstain from meddling which Scripture. Galileo apparently submitted. He undertook to put forward the Copernican theory merely as hypothesis, and for some time he adhered to the letter of his promise. But the irony, clever and biting, with which he wrote showed that an unsubdued hostility chafed within him. He produced a paper on the tides, which for him supplied the strongest argument in favor of the Copernican system, now, of course, known to be quite false. In 1618 three comets appeared, and in 1619 Galileo published his theory of comets, which he regarded as of similar nature with halos and rainbows. At the same time the Jesuit Grassi delivered a course of lectures, explaining them as heavenly bodies. It has been noted by one of his boigraphers that Galileo never, throughout his whole career, put forward a theory which was not remorselessly and bitterly attacked. This is not so astonishing when we consider with what ridicule and violence Galileo himself attacked Kepler's lunar theory of tides and Grassi's heavenlly theory of comets, bothe of which were, in fact, true, and with what animosity he carried his war into the camp of his opponents often without reason or justification. In 1624 Galileo again visited Rome and was cordially received, but failed to obtain, as he wished, a reversal of the judgment of the Inquisition. The new Pope, Urban VIII, had, as Cardinal Barberini, been one of Galileo's warmest friends and the astronomer seems to have expected that all would now go as he wished. He published an ironical discussion of the various theories of the universe, *"Il Saggiatore,"* which the Pope enjoyed, and, rendered over-bold by this success, brought out his "Dialogues" on the systems of the worlds (1632), in which, under the flimsiest veil, he boldly stated the case for the Copernican system. A supposed insult to the Pope, which Galileo certainly never intended, told against him. He was tried by the Inquisition, made a formal abjuration of his scientific theory, and retired to his villa at Arcetri, near Florence, a broken and disappointed old man. He died on Janary 8, 1642. A movement has recently been started in Rome to make the house at Arcetri, called "Jewel", a national monument, and to collect there all Galileo relics.

Gauss, Carl Friedrich. (1777-1855), German mathematician and astronomer who is generally regarded as one of the most prolific and influential mathematicians ever. In his doctoral thesis when he was only 22, he developed the concept of complex number and used it to establish the fundamental theorem of algebra. In 1801, he published *Disquisitiones arithmeticae,* which firmly established number theory as a well-defined branch of mathematics. He was Professor and Director of the observatory at Göttingen from 1807 onward, and was employed by the government to conduct a trigonometric survey of the kingdom of Hanover. He obtained a wide variety of essential results in geometry, algebra, analysis, astronomy, and statistics, as well as contributing to the mathematization of the physics of electricity, magnetism and gravitation. Gauss came of a peasant background and his extraordinary talent for mathematics showed itself at a very early age. By the age of three, he had doscovered for himself enough arithmetic to be able to correct his father's calculations when he heard him working out the wages for his laborers. Gauss retained a staggering ability for mental calculation and memorizing throughout his life. At the age of ten he astonished his schoolteacher by discovering for himself the formula for the sum of an arithmetical progression. As a result of such precocity the young Gauss obtained the generous patronage of the duke of Brunswick. The duke paid for Gauss to attend the Caroline College in Brunswick and the University of Göttingen, and continued to support him until his death in 1806. Gauss then accepted an offer of the directorship of the observatory at Göttingen. This post probably suited him better than a more usual university appointment since he had little enthusiasm for teaching. Working at the observatory no doubt also stimulated his interest in applied mathematics and astronomy. Gauss's life was uneventful. He remained director of the observatory for the rest of his life and indeed only rarely left Göttingen. Apart from mathematics he had a very keen interest in languages and at one stage hesitated between a career in mathematics and one in philology. His linguistic ability was evidently every great for he was able to teach himself fluent Russian in under two years. He also had a lively interest in world affairs, although in politics as in literature his views were somewhat conservative. Gauss's contributions to mathematics were profound and they have affected almost every area of mathematics and mathematical physics. In addition to being a brilliant and original theoretician he was a practical experimentalist and a very accurate observer. His influence was naturally very great, but it would have been very much greater had he published all his discoveries. Many of his major results had to be rediscovered by some of the best mathematicians of the 19th century, although the extent to which this was the case was only revealed after Gauss's death. To give but two of many examples — Janos Bolyai and Nikolai Lobachevsky are both known as the creators of non-Euclidean geometry, but their work had been anticipated by Gauss 30 years earlier. Cauchy's great poineering work in complex analysis is justly famous, yet Gauss had proved but not published the fundamental Cauchy theorem years before Cauchy reached it. The reason for Gauss's extreme reluctance to publish seems to have been the very high standard he set himself and he was unwilling to publish any work in a field unless he could present a complete and finished treatment of it. Gauss received his doctorate in 1799 from the University of Helmstedt for a proof of the fundamental theorem of algebra, i.e. the theorem that every equation of degree n with complex coefficients has at least one root that is complex number. This was the first genuine proof to be given; all the supposed previous proofs had contained errors, and it is this standard of rigor that really marks Gauss's work out from that of his predecessors. (mathematicians of the 18th century and earlier had often possessed an intuitive ability to conjecture mathematical theorems that were in fact true, but their ideas of rigorous mathematical proof fell short of modern standards.) Gauss's first publication is generally accepted as his finest single achievement. This is the *Disquisitiones Arithmeticae* of 1801. Appropriately it was dedicated to Gauss's patron the duke of Brunswick. The *Disquisitiones* is devoted to the area of mathematics that Gruss always considered to be the most beautiful, namely the theory of numbers or 'higher arithmetic.' Gauss's prodigious ability for mental calculation enabled him to arrive at many of his theorems by generalizing from large numbers of examples. Among many other striking results Gruss was able to prove in the *Disquisitiones* the impossibility of constructing a regular heptagon with straight edge and compass — a problem that had baffled geometers since antiquity. Gauss's intrest was not confined to pure mathematics and he made contributions to many areas of applied mathematics and mathematical physics. Thus he discovered the *Gaussian error curve* and also the method of least squares, which he used in his work on geodesy. In his work on electromagnetism he collaborated with Wilhelm Weber on studies that led to the invention of the electric telegraph. The invention of the bifilar magnetometer for his own experimental work was another practical consequence of Gauss's interest in electromagnetism. His interest in mathematical astronomy resulted in many valuable innovations; he obtained a formula for calculating parallax in 1799 and in 1808 he published a work on planetary motion. When in 1801 the asteroid Ceres was first observed and then 'lost' by Giuseppe Piazzi, Gauss was able to predict correctly where it would reappear. He also made improvements in the design of the astronomical instruments in use at his observatory. Gauss's work transformed mathematics and he is generally considered to be, with Newton and Archimedes, one of the greatest mathematicians of all time. The cgs unit of magnetic flux density is named in his honor.

Gelfand, Izrail Moiseyevich. (b. Okna [now Odessa Oblast], Russia, 2.9.1913). Russian mathematician. In the late 1930s he developed the theory of commutative normed rings (Banach algebras) which are of crucial importance in functional analysis and, increasingly, in modern physics. Then he turned his attention, with his collaborators, to the representation theory of locally compact groups, starting with the Lorentz group (the allowable transformations in special relativity theory). His work in classical Lie groups, ubiquitous in physics, with their analogues in algebraic geometry. He also developed integral geometry, which sutdies the transformation by integrals of functions on a given space in geometric terms. This technique has implications for his work on group representations, as does his third major area of interest, the cohomology of infimite dimensional Lie algebras. Gelfand's principal contributions are thus generalizing and extending classical mathematics by the astute but elegant use of infimite dimensional yet geometric ideas. For more than 20 years he was the dirctor of the Institute of Applied Mathematics of the USSR Academy of Sciences. Gelfand's mathematical ability revealed itself early, and althouth he had not completed the usual university education course his mathematical expertise was sufficient for him to be admitted to do postgraduate work at the Moscow State University at the age 19. He began teaching there in 1932 and was appointed an assistant professor in 1935. From 1939 Gelfand worked at the V. A. Steklov Institute of Mathematics of the Soviet Academy of Sciences, and since 1943 he has been professor at the Moscow State University. Gelfand is one of the most fertile and brilliant Soviet mathematicians of his generation, and his work has been equally influential in both pure and applied mathematics. His first major original contribution was made in his doctoral dissertation, presented in 1940, in which he made advances of great importance in the theory of commutative normed rings. Other outstanding achievements of Gelfand's include his contributions to group theroy, in particular his work on infinite dimensional representations of continuous groups, and on the harmonic analysis of noncompact groups. Gelfand carried out very important studies on the theory of generalized functions. This work focused on the application of such functions to be found in dealling with differential equations. As a result of this research the whole theory of intergral transformations was placed on a geometrical basis, and thus became amenable to a whole new range of techniques. In applied mathematics Gelfand's work has provided key mathematical tools needs in developing a theory of symmetry for elementary particles. From about 1958 Gelfand became interested in biology and physiology and he made studies of the nervous system and of cell biology. He applied mathematical techniques to these studies, for example devising mathematical models of neurophysiological systems.

Gellibrand, Henry. (b. Nov. 17, 1597; London; d. Feb. 16, 1636; London; English mathematician and astronomer. Gellibrand was educated at Oxford University. He was described by the diarist John Aubrey as, "good for little a great while, till at last it happened accidentally, that he heard a Geometrie lecture. He was so taken with it, that immediately he fell to studying it, and quickly made great progress in it." In 1662 he became Gresham Professor of Astronomy at Oxford. He did important work on evidence for the variation of the Earth's magnetic field publishing, in 1635, his observation of the 7^0 shift in direction of the compass needle over the previous 50 years.

Gell-Mann, Murray (b. Sept. 15, 1929; New York City) American thoretical physicist and mathematician. Gell-Menn graduated from Yale University in 1948 and gained his PhD from the Massachusetts Institute of Technology in 1951. He spent a year at the Institute of Advanced Study in Princeton, before joinging the Institute for Nuclear Studies at the University of Chicago, where he worked with Enrico Fermi. In 1955 he went to the California Institute of Technology, where he became a full professor in theoretical physics in 1956. Gell-Mann's chosen subject was the theoretical study of elementary particles. His first major contribution in 1953 (at the age of only 24) was to introduce the idea of 'strangeness'. The concept came from the fact that certain mesons were 'strange particles' in the sense that they had unexpectedly large lifetimes. This concept was also advanced independently by the Japanese physicist Kazuhiko Nishijima. Strangeness, as defined by Gell-Mann and Nishijima, is a quamtum property conserved in any so-called 'strong' interaction of elementary particles. The search for order among the known elementary particles led Gell-Menn and Israeli physicist Yuval Ne'eman to advance, independently, a mathematical representation for the classification of hadrons (particles that undergo strong interactions). Using group theory they showed that mesons and baryons could be classified into multiplets according to various properties. The mathematical group SU(3) — symmetry unitary theory of dimension 3 — is used. Particles are grouped into multiplets of 1, 8, 10, or 27 members and all particles in the same multiplet can be regarded as different states of the same basic particle. A major triumph for the theory was the prediction of the existence and properties of the omega-minus particle before it was observed at the Brookhaven National Laboratory in 1964. The theory was consolidated in Gell-Mann and Ne'eman's book *The Eighfold Way* (1964); the title refers to the octets of particles and, humorously, to the Buddhist eightfold path of morality designed to achieve nirvana. Gell-Mann felt that it should be possible to explain many of the properties of the known elementary particles by postulating even more basic particles, later to be called 'quarks'. (The name is from a quotation in *Finnegan's Wake* by James Joyce — "Three quarks for Master Marks.") These would have electric charges and baryon numbers that were integral numbers of one third. Quarks, together with their antiparticles, would normally be in combination as constituents of the more familiar nucleons and mesons. This idea challenged established thinking, and has greatly influenced the direction of highenergy theory and experiment. The quark hypothesis was further extended with the postulation of a fourth type of quark Sheldon Glashow (q. v.), and more recently other properties ('color') of quarks have been assigned. The unitary symmetry theory has been extended to a higher group SU(4). Gell-Mann received the 1969 Nobel Prize for physics, cited for his "contributions and discoveries concerning the elementary particles and their interactions." He is currently professor of theoretical physics at the California Institute of Technology's Lauritsen High-Energy Physics Laboratory.

Gosset William Sealy. (b. Canterbuty, Kent, UK, 1876; d. London, 16.10.1937). British statistician, writing as 'Student' who joined the Guinness strwerics in 1899 and worked for them is Dublin for three decades, being forced as a result to invent the technique of inference from small samples (1908-17). His first result is now known as the t-test for the population mean; it was the first parametric statistical test not to use the Normal distribution. 'Student' quickly added the t-test for the correlation coefficient from a small sample from a Normal population; and the technique of matched pairs in the design of experiments, in which correlation between two variables is deliberately used to reduce the sampling variation of their difference. These are still the first techniques taught to statisticians. L. McMullen & E. S. Pearson, 'William Sealy Gosset' *Biometrika,* 30 (1939), 205-50.

George Gamow. The Russian-American physicist George Gamow was born in Odessa, Russia, in 1904 and died in Boulder, Colorado, in 1968. Gamow's father was a literature teacher who instilled in his son a commitment to academic excellence which young George demonstrated at the Odessa Normal School from 1914 to 1920. Although the turbulent political conditions of Russia at that time interfered with Gamow's education, he enrolled in the University of Petrograd (Leningrad) where in 1925 he did optical experiments and studied relativistic cosmology under A. A. Priedmann. Although he was soon drawn to quantum theory and published his first paper on Schrödinger's wave equation in 1929, he continued to think about early cosmological models developed by Einstein, de Sitter and Lemaître. After receiving his Ph.D in 1928, Gamow went to Göttingen where he applied wave mechanics to alpha particles in a uranium nucleus to explain how these particles could cross a seemingly insurmountable nuclear energy barrier by tunneling through it. Gamow's work caused Niels Bohr to offer him a Carlsberg fellowship to spend a year at Bohr's Copenhagen Institute of Theoretical Physics where Gamow continued to study problems of nuclear physics. He also did original work on the interiors of stars. While on fellowship at the Cavendish Laboratory at Cambridge, Gamow estimated for Rutherford the amount of energy needed to split the nucleus using accelerated protons; Gamow's calculations prompted Rutherford to enlist J.D. Cockcroft and Ernest Walton to construct a successful particle accelerator. Gamow took a job at the University of Michigan after leaving the Soviet Union permanently in 1931. He became a professor of physics at George Washington University in 1934 where he collaborated with Edward Teller to develop the Gamow-Teller selection rule for beta decay. Gamow then began to apply the principles of nuclear physics to astronomical phenomena, developing a theory of stellar evolution based on nuclear reactions that ultimately contributed to Hans Beth's discovery of the carbon cycle. To understand the expanding Universe better, Gamow investigated the energy processes in red giant stars as well as the conditions causing stars to explode into novas and supernovas, hoping that these processes would yield a hint as to how the Universe began to expand. During World War II Gamow worked as a consultant with the Navy, studying the shock waves of various explosives. At the same time he continued to write a series of popular books; his first and perhaps most famous book is *Mr. Tompkins in Wonderland,* which contains amusing, nontechnical stories illustrating various principles of modern physics such as relativity theory. Gamow ultimately wrote (and in many cases illustrated) about thrity books, most of them for a lay audience. After the war Gamow began to construct a scenario for the first moments of the Universe's existence, suggesting that prior to the big bang there was primordial matter called "ylem" which consisted of neutrons, protons, and electrons jumbled together in a sea of high-energy radiation. Gamow suggested that these ingredients were responsible for the subsequent formation of the heavier elements once the Universe began to expand. The theoretical calculations of this plan appeared in a famous 1948 letter in *Physical Review* written by R. Alpher (who had done most of the calculations), Gamow, and Bethe, prompting Gamow to joke later that Bethe's name had been added to conform to the Greek alphabet. This letter later inspired a search for the pervasive cosmic radiation it predicted as the remnants of the original big bang explosion. This radiation was detected in 1964 by A. Penzias and R. Wilson, subsequent confirmations were provided by R. H. Dicke and P.J.E. Peebles.

George Sarton. George Sarton, who was born in Ghent, Belgium, in 1884, was possibly the most distinguished historian of science the world has ever seen. Almost single-handedly he developed his specialty into an independent discipline. He took his Ph.D. in mathematics at the University of Ghent, but he also studied chemistry and celestial mechanics. At the outbreak of world War I he left Belgium for England. In 1915 he emigrated to the United States. Before leaving Europe, he had founded *Isis*, an international quarterly review of the history of science, which first appeared in 1912. He continued publishing the journal in America and in 1936 he founded another periodical, *Osiris,* which was planned to include longer papers on the history and philosophy of science. From 1927 until 1947 he published three volumes entitled *Introduction to the History of Science,* covering the period from the time of Homer to the end of the fourteenth century. His plan for the complete history was to include nine volumes on the history of science through the nineteenth century. At the time of his death in 1956 he had published only the first two parts, *Ancient Science Through the Golden Age of Greece* and *Hellenistic Science and Culture in the Last Three Centuries* BC. He was fond of quoting a "theorem on the history of science" that he had written: Definition. Science is systematized positive knowledge, or what has been taken as such at different ages and in different places. *Theorem.* The acquisition and systematization of positive knowledge are the only human activities which are truly cumulative and progressive. *Corollary.* The history of science is the only history which can illustrate the progress of mankind.

Georg Riemann. Georg Fridrich Bernhard Riemann, one of the most original thinkers in the history of mathematics, was born in 1826 in Breselenz (Hanover), Germany, and died at the age of forty in Italy. His father was a Lutheran pastor, and Riemann originally intended to become a theologian. He learned Hebrew and made a valiant effort to demonstrate the validity of Genesis by using mathematics. He did not succeed in this effort, but the experience led him to abandon theology for mathematics. His university studies in Göttingen were interrupted by the revolutionary up-heavals of 1848, but with the victory of Kaiser Frederick Wilhelm IV over the radicals, Riemann returned to the university to finish his schooling. His thesis in mathematics met with the approval of the venerable mathematician Karl Gauss. Among the many importnat contributions made to mathematics by Riemann in his brief life, the most important is a version of non-Euclidean geometry presented to the public in 1854. The general public, insofar as it understood what Riemann was talking about, was shocked, for the brash nowcomer proposed a geometry in which the Euclidean axioms on paralles — that through a given point adjacent to a line, one and only one parallel line can be drawn—is abolishd in favor of a new axiom stating that through such a point no parallels can be drawn. Euclid had also taken as one of his axioms the principle that two points determine one and only one straight line; Riemann argued that any number of straight lines can be determined by such conditions. And finally, in Riemannian geometry, the sum of the angles of a triangle always exceeds 180^0. Riemann's geometry was shocking to those familiar with the ideal world of Euclid even though it is more appropriate to our own world experiences. As we live on the surface of a spherical planet, the shortest distance between two points is a segment of a great circle. Two perpendiculars or lines of longitude erected on the equator meet at the North or South Pole, thus creating a triangle whose base angles, where the longitudianl line and the equator meet, are each of 90^0; the third angle at the Pole, gives the polygon more than 180°. At best, Riemann's new geometry was considered a brilliant mathematical curiostiry, but fifty years after Riemann's death, Einstein demonstrated that his version of geometry fits reality more closely than did Euclid's. More specifically, Einstein in his general theory of relativity argued that space is positively curved due to the gravitational forces of the bodies distributed throughout the Universe. In Einstein's Universe, a line on an imaginary sphere surrounding a mass is not straight in the Euclidean sense but is, indeed, the shortest distance between two space-time points. Although the predictions of the general theory differ only slightly from those of Newton's theory, the latter's theory of gravitation suggested that the fact that two bodies travel around curved paths in each other's presence indicates an interacting gravitational field. Einstein, however, argued that what we perceive to be the gravitational attraction between two bodies is actually due to the geometric properties of space itself.

Gibbs, Josiah Williard. (b. Feb. 11, 1839; New Haven, Connecticut; d. Apr. 28, 1903; New Haven. American mathematician and theoretical physicist. Gibbs came from an academic family. He entered Yale in 1854, graduated in 1858, and in 1863 received a PhD for research on the design of gears. The same year he traveled to Europe, returning in 1869 to Yale where he remained until his death. In 1871 he was appointed professor of mathematical physics. His initial work on the theory of James Watt's steam-engine governor led him into a study of the thermodynamics of chemical systems. In a series of long papers published between 1873 and 1876 he developed, and indeed virtually completed, the theory of chemical thermodynamics. Gibbs's most famous paper, *On the Equilibrium of Heterogeneous Substances* (1876), contains the celebrated *Gibbs phase rule,* describing the equilibrium of heterogeneous systems. His name is also associated with the *Gibbs free energy* — a function that determines the conditions in which a chemical reaction will occur — and with several other equations in thermodynamics. Gibbs was also active in mathematics and physics. He worked on the theory of William Hamilton's quaternions and introduced the simpler, widely used, vector notation. Between 1882 and 1889 he published a series of papers on the electromagnetic theory of light. He also made important contributions to statistical mechanics, introducing the fundamental concept of *Gibbsian ensembles* — collections of large numbers of macroscopic systems with the same therodynamic properties, used in relating thermodynamic properties to statistical properties.Gibbs, who never married, lived a quiet retiring life at Yale; he was a poor teacher but a brilliant and productive theorist. His work, carried out far from the European mainstream of science, was largely published in the obscure *Transactions of the Connecticut Academy of Sciences.* However, James Clerk Maxwell understood the importance of his ideas as early as 1857 and in later life Gibbs was widely recognized. Many regard him as the greatest native-born American scientist.

Giorgio de Santillana and Hertha von Dechend. Giorgio de Santillana was for many years professor of the history of science at the Massachusetts Institute of Technology. He was a second-generation historian of science as his father had been one of the most distinguished practitioners of that discipline at the turn of the century.

De Santillana was thoroughly imbued with the culture of all civilizations and all arts and sciences. Hugh Trevor-Roper, the British historian, has written of him: His learning may range over vast tracts of time. He is sensitive to stimulus in many fields. He is at home in Babylon, in China, in Snerri Sturluson's Iceland, in untouched Polynesia, in per-Columbian America. He will quote Harbanus Maurus and Bertolt Brecht as effortlessly as Metrodorus or Athanasius Kircher, and carries with him, a little breathless perhaps, and dizzy, his throw-away illusions and polyglot versatility, from Anaximander and Parmenides to Einstein and Oppenheimer, from Hesiod and the epic Gylgamesh to Kafka, Auden, Salvemini, and Simone Weil. But three points of time and place touch him as he skims over them and bring him down to judge their dramatic quality and wrestle with their problems. These three points are the Greece—or rather the Greek dispersion—of the pre—Socratics, the Florence of Brunelleschi and Toscanelli, and Galileo's battle with Rome. Hamlet's Mill is said to lie at the bottom of the sea, where it grinds out the salt that makes the sea salty. In legend, it was known as Amlodhi's Quern, Amlodhi being the Danish original of Shakespeare's Hamlet. The selection by de Santillana and Hertha von Dechend is drawn from their whirlwind investigation of mythology, science, and time entitled *Hamlet's Mill.* It relies on materials similar to those explored by Emmanuel Velikovsky in his *Worlds in Collision.* Like Velikovsky, they find that the preliterate mythmakers were fearful of the specter of cosmic catastrophe. They base their conclusions on the inferences of long and precise astronomical observations and records. Unlike Velikovsky, however, our authors consider classical mechanics and astronomy along with their own interpretations of ancient myths in deducing what ancient disaster actually wrecked Hamlet's Mill.

Gödel, Kurt. (b. Brünn, Moravia, Austria-Hungary [now Brno, Czechoslovakia], 28.4.1906; naturalized American citizen, 1948; d. Princeton, NJ, USA, 14.1.1978). Austrian/ American mathematician and logician. Gödel worked at the University of Vienna from 1930 to 1940 when he emigrated to the USA. He settled at the Instituted for Advanced Studies at Princeton and remained there for the rest of his career. His most important work Über formal unentscheidbare Sätze' [On formally indeterminable propositions......] was published in 1932 (tr. J. van Heijenourt in *From Frege to Gödel,* Cambridge, Mass., 1966). In this he argued, with astonishing ingenuity but in an irresistibly coercive fashion, that in a certain strong and crucial sense mathematics is essentially incomplete. More specifically he proved that for any formal system that contains arithmetic there must be true statements of the system that cannot be proved within it. This completely under mined the 'logistic' project undertaken by Frege and Bertrand Russell of providing a set of logical axioms from which the whole of pure mathematics (as well as the nonaxiomatic remainder of logic) could be deduced. It also provided a brilliantly fruitful example of the kind of metamathematical inquiry — that is the strictly formal investigation of formal systems of logic and mathematics — to which logicians have productively applied themselves since the abandonment of the Frege-Russell project. He proved a number of fundamental metamathematical results that bear his name; in the course of these proofs he developed the theory of Recursive Functioning. and, since he thereby showed the unattainability of the aims of Hilberts Programme and (on some interpretations) Logicism, he brought about a complete reassessment of the Foundations of Mathematics. He also proved both the Axiom of choice and the Continuum Hypothesis to be consistent with the standard axioms of set theory.

Gödel, Kurt (b. Apr. 28, 1906; Brünn, now Brno in Czechoslovakia; d Jan. 14, 1978; Princeton, New Jersey). Austrian-American mathematician. Gödel initially studied physics at the University of Vienna, but his intersest soon turned to mathematics and mathematical logic. He obtained his Phd in 1930 and the same year joined the faculty at Vienna. He became a member, of the Institute for Advanced Study, Princeton, in 1938 and in 1940 emigrated to America. He was a professor at the Institute from 1953 to 1976, and received many scientific honors and awards including the National Medal of Science in 1975. In 1930 Gödel published his doctoral dissertation, the proof that first-order logic is complete — that is to say that every sentence of the language of first-order logic is provable or its negation is provable. The completeness of logical systems was then a concept of central importance owing to the various attempts that had been made to reveal a logical axiomatic basis for mathematics. Completeness can be thought of as ensuring that all logically valid statements that a formal (logical) system can produce can be proved from the axioms of the system, and that every invalid statement is disprovable. In 1931 Gödel presented his famous incompleteness proof for arithmetic. He showed that in any consistent formal system complicated enough to describe simple arithmetic there are propositions or statements that can neither be proved nor disproved on the basis of the aximos of the system — intuitively speaking, there are logical truths that cannot be proved within the system. Moreover, as a corollary Gödel showed (what is known as his second imcompleteness theorem) that the *consistency* of any formal system including arithmetic cannot be proved by methods formalizable within that system; consistency can only be proved by using a stronger system — whose own consistency has to be assumed. This latter result showed the impossibility of carrying out Hilbert's program (*see* David Hilbert), at least in its original form. Gödel's second great result concerned two important postulates of set theory, whose consistency mathematicians had been trying to prove since the turn of the century. Between 1938 and 1940 he showed that if the axioms of (restricted) set theory are consistent then they remain so upon the addition of the axiom of choice and the continuum hypothesis, and that these postulates cannot, therefore, be disproved by restricted set theory. (In 1963 Paul Cohen showed that they were independent of set theory.) Gödel has also worked on the construction of alternative universes that are models of the general theory of relativity, and has produced a rotatinguniverse model.

Gordan, Paul Albert (b. Apr. 27, 1837; Breslau, now in Poland; d. Dec. 21, 1912; Erlangen, now in West Germany). German mathematician. Gordan studied at Breslau, Königsberg, and Berlin and became professor of mathematics at the University of Erlangen. For most of his mathematical career his research was concentrated on a single field, the study of indeterminates. The central problem in the field, which Gordan eventually solved, was to prove the existence of a finite basis for binary forms of any given degree. His result was subsequently refined and extended by many workers including Gordan himself. Gordan's proof was long and complicated and the result was re-proved in 1888 by David Hilbert using newer and far simpler methods. In collaboration with Rudolf Clebsch, Gordan also wrote a book on Abelian functions that included the central theorem now know as the *Clebsch-Gordan theorem.* This work was influential in giving a new direction to algebraic geometry.

Gottfried Wilhelm von Leibniz. What might have been called the Battle of the Eighteenth Century had as protagonists Isaac Newton and his German counterpart, Baron Gottfried Wilhelm von Leibniz. Each man independently developed the calculus, that indispensable mathematical instrument without which modern physics and technology could never have developed. Each man was also the leading "natural philosopher" of his nation. Leibniz, four years younger than Newton, was born in 1646 in Leipzig. Germany, the son of a professor of moral philosophy at the University of Leipzig. He began reading in his father's library at the age of six. By the time he reached his fourteenth birhtday he was euipped with an excellent classical education. His achievements while still a young man in law, mathematics, and diplomacy were impressive. His services to the Elector of Hanover (who later became King George I of England) involved a great deal of travel abroad, so he became good friends with Oldenburg (who brought him into the Royal Society), Huygens, Spinoza, Viviani (a disciple of Galileo) and many others. He was one of the first great scientists to insist on the use of the vernacular instead of Latin and Greek in scientific writing, charging that many scholars used classical languages as a screen for concealing their ignorance. Just as people today who know nothing of astrophysics can heatedly argue the relative merits of the steady-state and big bang cosmologies, it did not take any great understanding of mathematics for people in all walks of life to take sides in the violent dispute which reged over who had first ivented the mysterious new mathematics of velocity and rates of change. The quarrel came to a head when a Fellow of the Royal Society named John Keill declared that Leibniz had actually passed off Newton's invension as his own. A committee was appointed to examine the propriety of Keill's accusation, but it went much further and published a report in 1712 saying that Leibniz had read an account of Newton's invention of his "method of fluxions" and published a variant of the method as his own creation. Newton's won role in the whole mess was far from laudable, involving the connivance of a Swiss named Facio de Duillier, an unsavory religious fanatic with whom Newton had at various times shared living quarters. After the publication of the committee's report, Newton attacked Leibniz at length in two successive issues of the *Philosophical Transactions* (January and February 1715) and is said to have told one of his students that he "had broke Leibniz's Heart with his Reply to him". The bitterness of the dispute between Newton and Leibniz over the discovery of calculus should not detract from the beauty and elegance of the method itself. Like arithmetic and algebra, calculus is a technique for reckoning. Arithmetic reckons with constant quantities represented by numerals. Algebra employs constant quantities but also has variable quantities, usually represented by letters (the variables t and x are typically used to designate time and space readings on a clock or a meterstick). Variable quantities are called variables for short. Calculus is form of algebra which includes infinitesimal quantities. Sence the time of Leibniz these quantities have been represented by compound symbols such as *dt* and *dx* (which are pronounced "dee tee" and "dee ex"). In differential calculus there is for every variable *x*, for example, a differential *dx*, which is thought of as a new independent variable representing a possible infinitesimal; change in x. The catch is that an infinitesimal quantity is one that is less than any fixed numerical quantity and in particular less than any fiexd fraction, a situation which drove Bishop Berkeley to accuse the physicists of placing greater demands on faith the Catholic Church and trafficking with the ghosts of departed quantities. Indeed some of the laws of arithmetic already known to Archimedes had to be suspended for infinitesimals and a logically consistent axiomatic treatment of infinitesimals had to wait until the twentieth century. Newton did not build his calculus on the differential in the same way as Leibniz. What Newton took as the primary new concept in his version of the differential calculus is the idea of the rate of change of one variable with respect to another variable such as time. Roughly speaking, if the variable x changes by ten when *t* changes by two, then the rate of change of x with respect to *t* is five. Newton would have written this rate of change as Dx while Leibniz would have expressed it as a quotient dx/dt ("dee ex by dee tee") of the two differentials *dx* and *dt*. Newton called this rate of change the fluxion of *x*, while the followers of Leibniz called it the differential quotient or differential coefficient. Either name would have been better than the term which finally stuck. By the twentieth century, the rate of change was almost uniformly called the "derivative" (of the one quantity with respect to the other). Special names were given to some derivatives: A derivative with respect to time is a rate, a derivative with respect to space is a gradient. The time derivative of position is velocity and the time derivative of velocity is acceleration. In an automobile, the odometer measures distance traveled and the speedometer measures speed. The odometer reading is one variable, the speedometer is another variable, and the clock is a third variable. The rate (or the time derivative) of the odometer reading is the speedometer reading. The other caclulus offered by Newton and Leibniz is the integral calculus. The concept of integral is a kind of inverse to the concept of rate. The reading of the odometer in an automobile is supposed to be the integral of the reading of the speedometer from when the car was new until the present. Integrals, like rates, arise when one quantity depends on another quantity. The integral of a variable like y between two values of another variable like t, called limits in this context, is the cumulative effect on y of the rate of change summed between the two time limits. An integral is thus a kind of summation and the symbol of the integral is a stylized S for the Latin *summa*. The limits of the integral are written below and above the integral sign. Leibniz and Newton also fought over the concept of *vis viva*, which was translated as "living force" and was supposed to be a single quantity representing the amount of motion in a body in some at first ill-defined way. Newton insisted that the *vis viva* of a body is its mass times its speed (mv) while Leibniz argued that it was its mass times the square of the speed (mv^2). Newton's candidate gives what we now call the momentum of a body, a very useful concept in physics. Leibniz's candidate, however, is indeed what we now call twice the kinetic energy of a body. As a result, Leibniz's conception survived as energy. It was subsequently renamed energy by Thomas Young and later split by William Rankine into potential and kinetic energy. When Leibniz sat down to write his *Monadology* he had already completed his important work in calculus and mechanics. He had seen the problems raised by Newton's absolute time and space and renounced them in favor of a relational theory of time and space Time and space, according to Leibniz, are the relations between successive and parallel happenings or events. He evidently regarded these happenings as more basic than either space or time, which was an importnat step toward the theory of relativity. Unlike his earlier work, however, Leibniz's *Monadology* ignored the experimental method almost entirely and soared off in a burst of aprioristic cosmological speculation, rejecting the dualistic compromise of atoms-in-a-plenum to search for a unifed plenum theory. He sought eternal and necessary truths and scorned mere empiricism. Leibniz assumed that the world was a collection of monads. As there was nothing else in the world, each monad had to have within itself complete information about the Universe as well as the ability to act upon this information as a kind of "divine machine or natural automation." As the *Monadology* had neither mathematical form nor connection with experiments, it was useless to physics. Yet it represented the first attempt in science to uncover a code of cosmic law rather than simply accept the effects of that law.

Gregory, James. (b. November 1638; Drumoak, Scotland; d. October 1675; Edinburgh). Scottish mathematician and astronomer. Gregory was one of the many 17th-century mathematicians who made important contributions to the development of the calculus, although some of his best work remained virtually unknown until long after his death. He studied mathematics at the University of Padua in about 1665 and produced *Vera Cireuli et hyperbolae quadratura* (1667; The True Areas of Circles and Hyperbolas). He was particularly interested in expressing functions as series, and he sketched the beginnings of a general theory. It was Gregory who first found series expressions for the trigonometric functions. He introduced the terms 'convergent' and 'divergent' for series, and was one of the first mathematicians to begin to grasp the difference between the two kinds. Gregory also gave the first proof of the fundamental theorem of calculus. In addition to his mathematical work Gregory's interests in astronomy led him to do some valuable practical work in optics. He anticipated Newton by recommending a reflecting telescope in his *Optica promota* (1663; The Advance of Optics). He realized that refracting telescopes would always be limited by aberrations of various kinds. His solution was to use a concave mirror that reflected (rather than a lens that refracted) to minimize these effects. He solved the problem of the observer by having a hole in the primary mirror through which the light could pass to the observer. However, he was unable to find anyone skilled enough actually of construct the telescope. Gregory held cahirs in mathematics at the University of St. Andrews (1669-74) and the University of Edinburgh (1674-75). He died at the age of 37 shortly after going blind.

Grothendieck, Alexandre (b. Berlin, Germany, 1928; naturalized French citizen). German/French mathematician. He arrived in France as a refugee in 1941, and for many years refused to take up French nationality out of respect for his Russian father, murdered by the Nazis. His major achievement marks a revolution in algebraic geometry. In collaboration with several other mathematicians, notably Bourbaki, he unified themes in geometry, number theory, topology and complex analysis. The crucial concept is the scheme, and the étale cohomology of schemes, developed in the 1960s' which proved capable of resolving the important number-theoretic Weil conjectures. The hardest conjecture was solved affirmatively by Grothendieck's pupil Pierre Deligne in 1972. Grothendieck's work also has implications for logic *via* his theory of topoi. He gave a purely theorem, and an algebraic definition of the fundamental group of a curve. His work is perhaps the most abstract significant development in mathematics since WW2. He was awarded a Fields medal in 1966. Grothendieck was prominently associated with an antimilitarist movement in the 1960s. He has turned recently to the teaching of mathematics, and since 1978 has been professor at Montpellier University. R. Hartshorne, *Algebraic Geometry* (NY 1977).

H

Hadamard, Jacques (b. Versailles, Yvelincs, France, 8.12.1865; d. Paris, 17.10.1963). French mathematicain. He first worked on complex function theory, where his theorem on lacunary series is fundamental to the detection of singularities of a function defined by its Taylor series. In 1896 he proved the prime number theorem, first raised by C.F. Gauss and notably discussed by G.F. B. Riemann: the number of prime numbers less than x is asymptotically equal to x/logax. This has been called the most important result ever obtained in number theory. It was also proved independently in the same year by de laVallcc Poussin. Hadamard's investigation of geodesics on surfaces of constant negative curvature (1898) initiated a study which continues today and has implications for probability theory, specifically in ergodic theory. It led Hadamard to the idea of well-or illposed problems in mathematical physics: a problems in well-posed as a differential equation if insignificant variations in the coefficients produce only insignificant variations in the solution, otherwise it is ill-posed because such behavior is unlikely to occur in reality. After WWI Hadamard's interests changed and his short book On *the Psychology of Invention in the Mathematical Field* (Princeton, 1945) provides many fascinating insights into the mathematical mind (notably J.H. Poincare's). Hadamard was an inspiring and influential lecturer at the Collège de France (1909-37), and was elected a member of the French Acadèmie des Sciences in 1912.

Hadley, John. (b. Apr. 16, 1682; England; d. Feb. 14, 1744; East Barnet, England). English mathematician and inventor. Little is known of Hadley's life. He is chiefly remembered for developing the reflecting telescope, producing his first in 1721. He was an extremely skilled craftsman and his reflectors were among the first to be useful in astronomy. Hadley also invented, in 1730, the reflecting quadrant with which a ship's position at sea could be determined by measurements of the sun or a star above the horizon. This instrument later developed into the sextant.

Hamilton, Sir William Rowan (b. Aug. 3/4, 1805; Dublin; d. Sept. 2, 1865; Dublin). lrish mathematician. Hamilton was a child prodigy, and not just in mathematics; he also managed to learn an extraordinary number of languages, some of them very obscure. In 1823 he entered Trinity College, Dublin, and four years later at the age of 22 was appointed professor of astronomy and Astronomer Royal for Ireland — posts given to his in order that he could continue to research unhampered by teaching commitments. In 1827 he produced his first original work, in the theory of optics, expounded in his paper. A theory of *Systems of Rays.* In 1832 he did further theoretical work on rays, and predicted conical refraction under certain conditions in biaxial crystals. This was soon confirmed experimentally. In dynamics he introduced *Hamiltion's equations* — set of equations (similar to equations of Joseph Lagrang) describing the positions and momenta of a collection of particles. The equations involve the *Hamiltonian function,* which is used extensively in quantum mechanics. *Hamilton's principle* is the principle that the integral with respect to time of the kinetic energy minus the potential energy of a system is a minimum. One of Hamiliton's most famous discoveries was that of *quaternions*. These are a generalization of complex numbers with for the property that the commutative law does not hold for them (i.e. A x B dies not equal B x A). Hamilton's discovery of such an algebraic system was important for the development of abstract algebra; for instance, the introduction of matrices. Hamilton spent the last 20 years of his life trying to apply quaternions to problems in applied mathematics, although the more limited theory of vector analysis of Josiah Willard Gibbs was eventually preferred. Toward the end of his life Hamilton drank increasingly, eventually dying of gout.

Hans Bethe. Hans Bethe, the discoverer of the nuclear processes that power stars, was born in Strasbourg, Alsace-Lorraine (then part of Germany), in 1906. He attended the University of Frankfurt for two years and then spent two and one-half years at the University of Munich, obtaining his doctorate in theoretical physics in 1928 under the direction of Arnold Sommerfeld. Briefly an instructor at Frankfurt and Stuttgart, Bethe became Privatdozent at Munich in 1930. He was working as an acting assistant professor at Tübingen in 1933 when he was fired by the Nazis. Faced with few prospects for work in Germany, Bethe emigrated to England and worked as a lecturer for a year and then spent a semester at the University of Bristol on a fellowship. He accepted an academic position at Corneu University in 1935 and became a full professor two years later. Except for sabbaticals and a prolonged absence during World War

II to work on mocrowave radar and to head the theoretical division at Los Alamos, Bethe has remained at Cornell where he is presently emeritus professor of physics. Before beginning his work in nuclear physics, Bethe devoted his attention to atomic physics. He also formulated a theory to explain inelastic collisions between atoms and fast particles, which provided a useful theoretical tool for nuclear physicists. Bethe also examined the splitting of atomic energy levels when an atom is inserted into a crystal, and worked as well on the theory of metals, developing a theory of the order and disorder in alloys. Bethe also laid much of the groundwork for the modern field of quantum electrodynamics by partially explaining the Lamb Retherford shift in the hydrogen spectrum. Bethe said that observed discrepancies between the energy levels of different states of the element were caused by the interaction of the electron with its own electromagnetic field. Bethe is best known for his contributions to the theory of atomic nuclei. In 1934 he and Rudolf Peierls developed the first theory of the deuteron. Bethe's work on nuclear reactions led to the discovery of the thermonuclear processes responsible for the creaction of energy in stars. Bethe hypothesized that most stars produce their energy by converting hydrogen into helium. His main contribution to the study of stellar energy processes was to exclude nuclear processes other than the carbon-nitrogen cycle, which is the most important nuclear reaction in very hot luminous stars, and the proton-proton reaction, which predominated in the sun and other cooler stars. Bethe was awarded the Nobel Prize in physics in 1967 for his work on stellar energy as well as other nuclear reactions. More recently, he has worked on a theory of nuclear matter which focuses on the forces between nucleons to explain the properties of atomic nuclei. Bethe has also become well-known for his opposition to government efforts to develop antiballistic missile systems and the Strategic Defense Initiative, so-called Star Wars intercontinental ballistic-missile shield.

Hans Reichenbach. Hans Reichenbach, the German-American philosopher of science, was born in Hamburg in 1891. He attended the Technische Hochschule at Stuttgart and studied at the universities of Berlin, Munich, and Göttingen before earning his doctorate in philosophy in 1915 from Erlangen. After teaching at Berlin from 1933 to 1938, followed by five years in Istanbul, Reichenbach came to the University of California at Los Angeles where he remained until his death in 1953. Although a logical positivist, he was not a member of the Vienna Circle. He did co-edit with Rudolf Carnap *Erkenntnis,* which was later renamed *The Journal of Unified Science.* Reichenbach made important contributions to the study of probability, quantum mechanics, space, time, relativity, geometry, and scientific laws. Reichenbach's eassy suggests that relativity theory is both physical and philosophical in its implications. It is physical because it describes space-time in terms of geometry; it is philosophical because it has altered the way we view space and time and has focused our attention on the inability of the language of classical physics to describe the phenomena of a four-dimensional relativistic Universe. Reichenbach argues that a physical philosophy usually results from a breakthrough in physical theory. A science philosopher views a physical theory in terms of the logical relations that establish its validity and its progression from theory to fact Like Heisenberg, Reichenbach touches upon the problem of the lack of precision in language. He adds that Einstein's theory of relativity is the product of connections of different space-time description in which alternative words may be used to describe the same events. Reichenbach's essay also grapples with our intuitive sense that space and time are distinct from each other. Although the concepts of space and motion evolved over a long period of time, there was no similar evolution in our perception or concept of time. Even the Austrian physicist Ernst Mach spoke of relativity as a measure of uniform time. This view of an unchanging time axis did not give way until Einstein advanced his theory about the relativity of simultaneity, which had originated in the failure of the Michelson-Morley experiment to detect the other medium. Although time is not theoretically unidirectional in relativity theory, it is asymmetrical since, otherwise, it would be impossible to maintain an orderly existence in the real world. Yet there is nothing in Einstein's equations which states that time is irreversible. Perhaps Einstein's most significant contribution to the study of time was to show it has no central axis but instead varies with changing frames of reference. In this way Einstein avoided the absolutism of Newton and the need to formulate hypotheses that assume some universal time or place.

Hardy, Godfrey Harold. (b. Feb. 7, 1877; Cranleigh, England; d. Dec. 1, 1947; Cambridge, England). British mathematician. Hardy had his mathematical education at Cambridge University and remained there as a fellow of Trinity College until 1919 when he became Savilian Professor of Geometry at Oxford. From 1931 to 1942 he was back in Cambridge as Sadleirian Professor of Pure Mathematics. His central field of interest was in analysis and such related areas as convergence and number theory. *Hardy classes* of complex functions are named for him. For 35 years, starting in 1911, Hardy collaborated with J.E. Littlewood and together they wrote nearly a hundred papers. The principal areas they covered were Diophantine approximations, the theory of numbers, inequalities, series and definite integrals, and the Riemann zeta-function. Although primarily a pure mathematician Hardy made one lasting contribution to applied mathematics; the *Hardy—Weinberg law* was discovered independently by Hardy and the physician Wilhelm Weinberg in 1908 and proved to be fundamental to the science of population genetics. It gives a mathematical description of the genetic equilibrium in a large randommating population and explains the surprising fact that, unless there are outside changing forces, the proportion of dominant to recessive genes tends *not* to vary from generation to generation. The law offered strong confirmation for the Darwinian theory of natural selection. Hardy was one of the outstanding British mathematicians of his day, an excellent teacher, and one of the first to introduce modern work on the rigorous presentation of analysis into Britain. His *Course of Pure Mathematics* (1908) was influential on the teaching of mathematics in British universities. One of his achievements was his discovery of the young Indian mathematician Srinivasa Ramanujan. Partly through Hardy's efforts Trinity College made funds available for Ramanujan to go to Cambridge to pursue his mathematical researches under Hardy. Hardy was a passionate devotee of cricket and an equally passionate enemy of the Christian religion. During World War I he was a staunch supporter of Bertrand Russell when Trinity set about depriving Russell of his position on account of his pacifist activities.

Hardy wrote a lively autobiographical sketch. *A Mathematician's Apology.*

Hawking, Stephen William. (b. Jan. 8, 1942; Oxford, England). British theoretical physicist. Hawking graduated from Oxford University and obtained his PhD from Combridge University. After being connected with various Cambridge institutes and departments, he was appointed in 1977 to the chair of gravitational physics. Hawking has worked mainly in the field of general relativity and in particular on the theory of black holes. He has objected to Einstein's treatment of gravity in his general theory since "it treats the gravitational field in a purely classical manner when all other observed fields seem to be quantized." A further objection, made with G.F.R. Ellis in their *Large Scale Structure of Space Time* (1973), was that the theory led inevitably to singularities that it could not describe adequately. Two such singularities they suggest are the totally collapsed form of stars known as black holes and the big bang beginning the expansion of the universe. For these reasons Hawking has been one of the leaders in the search for a theory of quantum gravity. Such a search has yet to find may general success; although a number of theories have been advanced it is quite clear that all still possess serious defects. In the theory of black holes Hawking has been more successful, establishing a number of remarkable theorems. Black holes are celestial 'bodies' that having had a mass in excess of three solar masses have undergone a gravitational collapse so extreme that they contract below the critical radius, calculated by Karl Schwarzschild, at which light or any other signal can escape. At first it appeared that absolutely nothing could be known about such bodies, or singularities, but Hawking has managed to construct many of their properties and show their relationship to move classical parts of physics. Hawking showed however that black holes could originate in other circumstances. There could be "a number of very much smaller black holes scattered around the universe, formed not by the collapse of stars but by the collapse of highly compressed regions.......that are believed to have existed shortly after the 'big bang' in which the universe originated." These 'mini black holes' could weigh a billion tons and yet be no bigger than a proton with a radius of 10^{15} meters. His mostexciting result, published in 1974, was one that he confessed he found hard to believe. This was the claim that black holes are not 'black' but emit particles at a steady rate. This result has been repeatedly confirmed mathematically and Hawking is able to propose a physical quantum process that would produce the effect. Quantum mechanics supposes space to be full of 'virtual' particles, i.e. particles that cannot be observed but do exist. They exist as pairs of particles and antiparticles that are constantly joining, separating, and annihilating each other. If one member of the pair were to be attracted into a black hole leaving its partner alone, then 'The forsaken particle or antiparticle may fall into the black hole after its partner but it may also escape to infinity, where it appears to be radiation emitted by the black hole." Such conclusions emerged when Hawking and his colleagues were able to link the physics of black holes with the laws of thermodynamics. It should be said that since the early 1960s Hawking has been the victim of a progressive nervous disease. This has confined him to a wheelchair and has prevented him from writing or calculating in a direct and simple way.

The bulk of his work, involving complex calculations, difficult mathematical proofs, and the introduction of new physical ideas, is thus interwoven into presentable form purely in his mind.

Heaviside, Oliver. (b. London. U.K. 18.5. 1850; d. Torquay, Devon, 3.2.1925). Heaviside was one of those men of genius who work best alone. By the age of 20 he was using the operator j A -1 as a standard piece of algebra needed to solve acnetworks, even though ac did not come into common use for a further 15 years. He coined the words 'inductance', 'impedance' and 'attenuation' now so familiar to all engineers. He went on to deal with the transient behaviour of net- works using another operator p which was tantamount to taking a differential, though not precisely so. Soon he had a meaning for A p. With such revolutionary concepts it is hardly surprising that his manuscripts submitted for publication were rejected. He received very harsh treatment from his peers and this left him embittered. He declared his main research completed by 1887, and 10 years later took on himself the life of a hermit. He was the first man to suggest that a single telophone line could convey more than one message at a time. He suggested that a conducting layer in the upper atmosphere (now known as the 'Heaviside layer') might act like a conducting sheet while the sea provided another, and that radio propagation was guided and reflected between the two as between transmission lines. This was some years before Appleton demonstrated the existence of the layer by experiment. Heaviside suggested the use of inductance-loading in telephone lines to reduce acoustic distortion. At 23 he read Maxwell's work extensively and ever after declared himself the apostle of Maxwell. 'I then put Maxwell aside and followed my own way. And I advanced much faster'. During this period he established a symmetry for the whole of electromagnetics. In his main published work *Electromagnetic Theory* (3 vols, London, 1893-1912) he wrote: 'As the universe is boundless one way towards the great, so it is equally boundless the other way towards the small, and important events may arise from what is going on inside atoms and again in the inside of electrons..... From the atom to the electron is a great step. But it is not finality' — this only two years after the discovery of the electron. He had assembled a fourth volume of this work, but it was torn up and scattered by burglars a few days after his death. From partial reconstruction it is clear that it contained a unified field theory to embrace both gravitation and electromagnetism: he called 'twisted nothingness' what Einstein later termed 'curved space'. In 1950 Professor Bjerknes wrote: 'I proposed Heaviside for the Nobel Prize, but alas it was 100 years too early.'

Hermann Bondi. Hermann Bondi was born in Vienna in 1919 and came to England as a young man to study at Cambridge University. In 1954 he left Cambridge to take the chair in mathematics at King's College, London. In addition to his academic career, Bondi also served as the chief scientific adviser to the Ministry of Defence and chief scientist at the Department of Energy. He was knighted in 1973 in recognition of both his scientific work and his contributions to public service. Bondi is currently the Master of Churchill College at Cambridge. Bondi's work has centered on applied mathematics and cosmology. He is best-known for his 1948 proposal with the astronomer Thomas Gold of the steady-state theory of creation. Bondi and Gold's theory suggested that the large-scale uniformity of the Universe as well as its outward expansion could be reconciled if one allowed for the continuous creation of matter. The steady-state theory was contradicted by the discovery by Penzias and Wilson in 1964 of the microwave radiation thought to be the echo of the big bang. Earlier inconsistencies of the steady-state theory had prompted a number of astronomers to devise variants of the general continuous creation theme to agree better with astronomical observations. The British astrophysicist W.H.McCrea, for example, proposed that new matter is created not from empty space but by existing matter. He suggested that all galaxies give rise to the creation of matter and that a new galaxy begins to form when an older galaxy becomes too large and ejects a fragment of itself. Fred Hoyle, the British cosmologist, and J.V. Narlikar, the Indian astrophysicist, tried to explain the act of creation with a negative energy reservoir that, they postulated, creates all matter; the more matter created, the more intense this energy reservoir becomes. Although Hoyle and Narlikar and their colleagues showed admirable creativity in trying to shore up the steady-state theory, they were less than successful in explaining the presence of heavier elements in the Universe supposedly formed during the big bang explosion or in explaining the detection in 1964 of the 3° K isotropic radiation in space that is believed to be the "ehco" of the primordial explosion. Bondi focuses on the attractive aspects of the steady-state theory, wondering why the physical processes we observe on earth are not common throughout the Universe. The aesthetic appeal of the theory is undeniable; it is the only cosmological scheme that is symmetrical both in space and time. Although the creation of matter provides a means for renewing the Universe without requiring it to reemerge phoenix-like from a gravitationally collapsed state, it is a theory that relies as much on other' disproving it as it does on its own proponents. And it requires the creation of one hydrogen atom in the space of a living room every few million years, a rate far too slow to be detected. Today the steady-state theory can boast only a few adherents as cosmologists have focused their efforts on reconstructing the big bang.

Hermann Helmholtz. Hermann Helmholtz was born in Potsdam, Germany, in 1821 and died in Berlin in 1894. Like the sons of many German schoolteachers young Hermann did what his father wanted him to do, not what he himself preferred. He wanted to a physicist but his father sent him to medical school, where he was eligible for aid from the Ministry of War that would not have been available to an aspiring physicist. In return for this, Helmholtz served for a number of years as an army doctor. In 1849 a friend wangled a position for him as lecturer in anatomy at the Academy of Arts in Berlin. He went on to become a professor of physiology at the University of Königsberg; in 1858 he became professor of anatomy at Heidelberg, and finally, in 1871, he started teaching physics in Berlin. An ophthalmologist will use a little flashlight to look into the interior of the eye. This instrument was invented by Helmholtz, who also invented the ophthalmometer, an instrument that measures the curvature of the eye. In his double capacity as physicist and physiologist he took up the work of another physician-physicist, Thomas Young, and revised Young's theory of three-color vision, with sufficient success that what had been called the Young theory became known as the Young-Helmholtz theory. Helmholtz did splendid pioneering work on the physiology of the ear and the physics of sound. He proposed the theory that differences in pitch are detected by the cochlea, a spiral organ in the inner ear equipped with progressively smaller resonators that detect progressively higher frequencies. He also discovered the role of overtones in determining tone quality and analyzed the phenomena of accordance and discordance—that is, whether a combination of simultaneously sounded notes sounds "good" or "bad." One of Helmholtz's teachers, a certain Müller, taught that the speed of a nerve impulse could never be measured because it moved so fast through so short a distance. But Helmholtz was stubborn and ultimately did derive a method of successfully measuring the speed of nerve impulses; his schoolteacher father must have been pleased with his tenacity. Helmholtz extended the principle of the conservation of energy to include all known forms of energy. He tried to explain the tremendous output of the sun by theorizing that the source of its heat was the gravitational condensation of the sun from its own diffuse primordial gases. But this explanation allowed for a solar lifetime of only a few million years, a timespan far too short to account for the evolution of life on earth. The suggestion that the sun is powered by a process of thermonuclear fusion had to wait until the noted physicist Hans Bethe published his famous paper in a 1936 issue of *Physical Review*. As was mentioned above, Helmholtz independently discovered the hypothesis of the conservation of energy and presented it in so elegant a form that for a ong time he was credited as the sole discoverer. Our selection from Helmholtz, instead of reading like a detailed report on a highly specialized theory, is more a rambling examination of a number of phenomena culminating in a brief reference to the conservation of energy, with ample credit being given to Mayer and others.

Hermite, Charles. (b. Dec. 24, 1822; Dieuze, France; d. Jan. 14, 1901; Paris). French mathematician. Hermite's mathematical career was almost thwarted in his student days, since he was incapable of passing exams. Fortunately his talent had already been recognized and his examiners eventually let him scrape through. Hermite obtained a post at the Sorbonne where he was an influential teacher. Hermite began his mathematical career with pioneering work on the theory of Abelian and transcendental functions, and he later used the theory of elliptic functions to give a solution of the general equations of the fifth degree — the quintic. One longstanding problem solved was proving that the number '*e*' is transcendental (i.e. not a solution of a polynomial equation). He also introduced the techniques of analysis into number theory. His most famous work is in algebra, in the theory of *Hermite polynomials.* Although Hermite himself had little interest in applied mathematics this work turned out to be of great use in quantum mechanics. He developed the theory of functions, used Elliptic Functions to solve the general quintic in one variable, and proved that *e* is Transcendental. Although he had already done original mathematical work, he found examinations difficult and it took him six years to obtain his first degree.

Hero of Alexandria. (*fl.* AD 62). Greek mathematician and inventor. Hero produced several written works on geometry, giving formulae for the areas and volumes of polygons and conics. His formula for the area of a triangle was contained in *Metrica,* a work that was lost until 1896. This book also describes a method for finding the square root of a number, a method now used in computers, but known to the Babylonians in 2000 BC. In another of Hero's books, *Pneumatica,* he wrote on siphons, a coin-operated machine, and the aeolipile—a prototype steam-powered engine that he had built. The engine consisted of a globe with two nozzles positioned so that steam jets from the inside made it turn on its axis. Hero also wrote on land-surveying and he designed war engines based on the ideas of Ctesibius. Yet another of his works, *Mechanica,* was quoted by Pappus of Alexandria.

Hilbert, David. (b. Königsberg, East Prussia, Germany [now Kaliningrad USSR]. 23.1.1862; d. Göttingen, Lower Saxony, Germany. 14.2.1943). German mathematician, educated and first employed at Königsberg, removing to Göttingen in 1985. Hilbert started by effectively terminating the existing theory of invariants. Essentially he did so by solving a more general class of problems and so being led to initiate modern algebraic number theory. Perhaps the most pervasive single concept in modern mathematics is the Hilbert space. It can be regarded as a generalization of ordinary n-dimensional space in which n is allowed to go to infinity and the concept of distance is preserved by insisting that the sum of the squares of coordinates be a converging sequence. Hilbert was led to this in developing functional analysis; if you wish to give a standardized description of a function that is based on an orthonormal sequence of standard functions (its spectrum) that description will lie in a Hilbert space. This idea has been used in abstract mathematics, in electrical engineering, in quantum physics, and in many other contexts. Hilbert decisively increased the prestige of the axiomatic method in mathematics. In his *Grundlagen der Geometrie* [Foundations of geometry] (Leipzig. 1899,[9] 1962) his method was to define his subject completely in terms of a finite set of axioms, all mutually consistent and none redundant. He could then use various algebraic models of the resulting structure to prove results without appealing to intuition. For the rest of his life Hilbert propounded the view that all mathematics should be handled on these principles, becoming a network of structures whose existence is strongly independent of the mathematicians who create them. This programme has been shown to be impossible (Gödel) and even unthinkable (P.J. Cohen), as well as undesirable (L.E.J. Brouwer). But it has resulted in the normative approach among modern mathematicians; the most influential event in 20c mathematics was Hilbert's proposal of 23 problems, conceived in this spirit (1900). Most have now been solved; each solution was a major event; and many practitioners have been brought up to see mathematics as primarily a discrete set of challenges. He is best known for his influential work in the foundations of geometry and mathematics in general; Hilbert's Programme was the motivation for the development of Computability Theory. His collection of 23 problems, now referred to as Hilbert's Problems, profoundly influenced the course of 20th-century mathematics; these problems, many of which still unsolved, are listed in Appendix 3. His other achievements included Hilbert's Basis Theorem of Ring theory, and his investigations of Hilbert Space theory and number theory.

Hilbert, David. (b. Jan. 23, 1862). Königsberg, now kaliningrad in the Soviet Union; d. Feb. 14, 1943; Göttingen, now in West Germany). German mathematician. Hilbert studied at the universities of Königsberg and Heidelberg and also spent brief periods in Paris and Leipzig. He took his PhD in 1885, the text year became *Privatdozent* at

Königsberg, and by 1892 had become professor there. In 1895 he moved to Göttingen to take up the chair that he occupied until his official retirement in 1930. Hilbert's mathematical work was very wide ranging and during his long life there were few fields to which he did not make some contribution and many he completely transformed. His attention was first turned to the newly created theory of invariants and in the period 1885-88 he virtually completed the subject by solving all the central problems. However his work on invariants was very fruitful as he created entirely new methods for tackling problems, in the context of much wider general theory. The fruit of this work consisted of many new and fundamental theorems in algebra and in particular in the theory of polynomial rings. Much of his work on invariants turned out later to have important application in the new subject of homological algebra. Hilbert now turned to algebraic number theory where he did what is probably his finest research. Hilbert and Minkowski had been asked to prepare a report surveying the current state of number theory but Minkowski soon dropped out leaving Hilbert to produce not only a masterly account but also a substantial body of original fundamental new discoveries. The work was presented in the *Zahlbericht* (1897) with an elegance and lucidity of exposition that has rarely been equalled. Hilbert then moved to another area of mathematics and wrote the *Grundlagen der Geometrie* (1899; Foundations of Geometry), giving an account of geometry as it had developed through the 19th century. Here his interest lay chiefly in expounding and illuminating the work of others in a systematic way rather than in making new developments of the subject. He devised an abstract axiomatic system that could admit many different geometries — Euclidean and non-

Euclidean — as models and by this means go much further than had previously been done in obtaining consistency and independence proofs for various sets of geometrical axioms. Apart from its importance for pure geometry his work led to the development of a number of new algebraic concepts and was particularly important to Hilbert himself because his experience with the axiomatic method and his interest in consistency proofs shaped his approach to mathematical logic and the foundations of mathematics. In mathematical logic and the philosophy of mathematics Hilbert is a key figure, being one of the major proponents of the formalist view, which he expounded with much greater precision than had his 19th century precursors. This philosophical view of mathematics had a formative impact on the development of mathematical logic because of the central role it gave to the formalization of mathematics into axiomatic systems and the study of their properties by metamathematical means. Hilbert aimed at formalizing as much of mathematics as possible and finding consistency proofs for the resulting formal systems. It was soon shown by Kurt Gödel (q.v.) that *Hilbert's program,* as this proposal is called, could not be carried out, at least in its original form, but it is none the less true that Gödel's own revolutionary mathematical work would have been inconceivable without Hilbert. Hilbert's contribution to mathematical logic was important, especially to the development of proof theory, as further developed by such mathematicians as Gerhard Gentzen. Hilbert also made notable contributions to analysis, to the calculus of variations and to mathematical physics. His work on operators and on *Hilbert space* (a type of infinite-dimensional space) was of crucial importance to quantum mechanics. His considerable influence on mathematical physics was also exerted through his collegues at Göttingen who included Minkowski, Hermann Weyl, Erwin Schrödinger, and Werner Heisenberg. In 1900 Hilbert presented a list of 23 outstanding unsolved mathematical problems to the International Congress of Mathematicians in Paris. A number of these problems still remain unsolved and the mathematics that has been created in solving the others has fully vindicated his deep insight into his subject. Hilbert was an excellent teacher and during his time at Göttingen continued the tradition begun in the 19th century and built the university into an outstanding center of mathematical research, which it remained unitil the dispersal of the intellectual community by the Nazis in 1933. Hilbert is generally 20th century and indeed of all time.

Hill, Austin Bradford. (b. London, U.K, 8.7. 1897). British medical statistician. Professor of medical statistics at London University. Hill was responsible for introducing into medicine the concept of the controlled clinical trial. Hill showed that when a new treatment became available it was best tested by comparing two groups of patients, otherwise as similar as poosible, one group being given the drug and the other a placebo or dummy treatment. He also applied statistical methods to other medical problems and (with Sir Richard Doll) showed the connection between smoking and lung cancer in a large-scale study of British doctors in the 1950s.

Hipparchus. About the middle of the third century B.C., a spirit of realism entered into Greek thought, and one of the results was that astronomy began to be treated as a science. That is to say, philosophers were no longer satisfied merely to produce fanciful theories of the universe based upon conjecture rather than upon fact. Observation and calculation became predominant, and hypotheses were relegated to their rightly subordinate place. The systems of Eudoxus and of Aristarchus alike broke down under the test of more careful and detailed observation of the planetary motions. It is the great merit of Hipparchus, the famous predecessor of Ptolemy, that he was more interested to investigate the actual motions and irregularities of the planets than to build up an elaborate theory of the universe. Hipparchus was born at Nicaca, in Bithynia, nearly at the beginning of the second century before our era. Nothing is known of his personal history. The earliest recorded date of his observations is 161 B.C., and his only book which has come down to us bears the date of 140 B.C. A large part of his life was spent at Rhodes, which was a center of great intellectual activity. Here he made his observations and, in 130, the capital discovery of the precession of the equinoxes. The outburst of a new state in 134 is stated by Pliny to have suggested the preparation of his catalogue of 1080 stars divided into six classes of magnitude, which is one of the finest monuments of ancienct astronomy, and substantially embodied in Ptolemy's *"Almagest"*. Hipparchus founded trigonmetry and compiled the first table of chords. This is known to us through the words of Theon on Alexandria. Scientific geography originated with his invention of the method of fixing terrestrial positions by circles of latitude and longitude. Only the inventor of the astrolabe. He was one of his many works has survived, a commentary on the *Phenomena* of Aratus and Eudoxus. There can be little doubt that the fundamental part of his astronomical knowledge was derived from Chaldea, but his borrowings were doubtless independently verified. It is probable, too, that he is the hipparchus who wrote on combinatory analysis, and that the Arabs were correct in atributing to him a knowledge of the quadratic equation. The system of Hipparchus, which was adopted and elaborated by Ptolemy, was in many points erroneous. None the less, it marked an immense advance, inasmuch as it was really a serious attempt to describe in reasoned form the actually observed motions of the plantes. It contains, therefore, a substratum of truth, and has been of inestimable value as a foundation of astronomical science. Hipparchus worked out his system most fully in relation to the sun, and showed that the supposed movements of that body could be accounted for either on the theory of the eccentric circle or on that of the epicycle. This was reasonable enough, for the resultant error was too small to be appreciable with the instruments which astronomers had then at their command. Hipparchus was also the discoverer of the precession of the equinoxes—that is to say, of the fact that the nodes, or intersections, of the earth's orbit, which was then, of course, regarded as the sun's orbit, with the plane of the equator were not absolutely fixed, but had a slow backward movement. This ancient astronomer also worked out a theory of the moon and her elusive movements, but with less success than that which attended his explanations of the sun. It is characteristic of Hipparchus that he realized the imperfections of his theory, though he confessed himself unable to remove them. In a similar way, though in a much greater degree, he was to the end dissatisfied with his knowledge of the movements of the planets — so much so that he never elaborated a planetary theory. He corrected the observations, which had been made by his predecessors, and was himself a scrupulously accurate observer, but he deliberately left the interpretation of these results to future generations, In this respect, as in so many others, we can see that his views of human knowledge, and of the celestial system alike, were far more profound than those which were current in his time. Early Greek astronomers had been accustomed to found their theories upon observation of the planets at their moments of opposition only, but Hipparchus, on the contrary, recognized that the movements of these bodies must be sedulously followed throughout their entire orbits. In making through observations of this kind, he soon found that neither the epicycle nor the eccentric was capable of explaining planetary motion, and there he was obliged to leave the matter — to Copernicus. In addition to his work on the solar system, Hipparchus is known to have complied a catalogue of stars, which formed the basis of Ptolemy's later catalogue. Ptolemy, who fixed its date at 129 B.C., did very little more than edit it.

Hodge, Sir William Vallance Douglas. (b. June 17, 1903; Edinburgh; d. July 7, 1975; Cambridge, England). British mathematician. Hodge studied at the universities of Endinburgh and Cambridge, then taught at Bristol University (from 1926) and in America at Princeton (1931-32). Most of his career was spend in Cambridge, England, where in 1936 he took up the Lowndean Chair in Mathematics. Hodge's mathematicalwork belongs almost entirely to algebraic geometry and although no expert on analysis or physics his work had immense impact in both these fields. Hodge's principal contribution was to the theory of harmonic forms. One of his central results was a uniqueness theorem showing that there is a unique harmonic form with prescribed periods. In general Hodge helped to initiate the shift of focus in mathematics from a search for purely local results to the more ambitious global approach now so influential.

Horrocks, Jeremiah. (b. *c.* 1617; Toxteth, near Liverpool, England; d. Jan. 3, 1641; Toxteth). English astronomer and mathematician. Horrocks studied at Cambridge University from 1632 to 1635 and became curate of Hoole in Lancashire in 1639, studying and practicing astronomy in his spare time. He very quickly saw the importance of Johannes Kepler's ideas, which was by no means common in the early 17th century. Using Kepler's Rudolphine tables he became the first man both to predict accurately and to observe a transit of Venus across the face of the Sun. He realized the significance of this relatively rare event and set about using it to determine the solar parallax (distance) and diameter. Unfortunately these results, together with his defense of Kepler, were only published in 1673, many years after his early death.

Hounsfield, Godfrey Newbold. (b. Newark, Notts, UK. 28.8. 1919). British inventor. A research scientist at EMI. Hounsfield developed the first computerized axial scanner (CAT scanner). This X-ray apparatus produces far clearer pictures of the brain and of the internal organs than conventional X-ray films. As Hounsfield's scanner rotates around the head, a beam of X-rays passes through from each of hundreds of points around the circle. The X-rays are received by a detector which calculates the proportion of X-rays absorbed by the bone, muscle and other tissue in their path. This information is then used by the computer built into the machine to create a cross-sectional picture of the head. The same process can be repeated at intervals of a fraction of an inch to produce a whole series of cross sections. Modern scanners can be used to examine any part of the body, and with increasing computer sophistication the date can be used to produce sectional pictures of the body in any plane. By comparison with conventional techniques, CAT scanners require a very low dosage of X-rays and without any of the discomfort unavoidable with techniques, such as angiography, which require injection of dyes into the bloodstream. Hounsfield was educated in Nottinghamshire and went on to the City and Guilds College, London, and the Faraday House College, Electrical Engineering in London. He worked for Electrical and Musical Industries (EMI) from 1951 and led the design effort for Britain's first large solid-state computer. Later he worked on problems of pattern recognition. He is now head of the medical research division of EMI. Although he had to formal university education he was granted an- honorary doctorate in medicine by the City University, London (1975). Hounsfield was awarded the 1979 Nobel Prize for medicine, together with the South-African-born physicist Allan Cormack, for his pioneering work on the application of computer techniques to x-ray examination of the human body. Working at the Central Research Laboratories of EMI he developed the first commercially successful machines to use computer-assisted tomography, also known as computerized axial tomography (CAT). In CAT, high-resolution x-ray picture of an imaginary slice through the body (or head) is built up from information taken from detectors rotating around the patient. These 'scanners' allow delineation of very small changes in tissue density. Introduced in 1973, early machines were used to overcome obstacles in the diagnosis of diseases of the brain, but the technique has now been extended to the whole body. Although Cormack worked on essentially the same problems of CAT, the two men did not collaborate, or even meet. New lines of research being pursued by Hounsfield include the possible use of nuclear magnetic resonance (NMR) as a diagnostic imagine technique.

Huygens, Christian. After the death of Galileo and of Kepler, and before Sir Isaac Newton rose to fame, there is an interval during which Huygens stands out as by far the greatest astronomer of his day. Born at the Hague on April 14, 1629, Christian Huygens de Zuylichem belonged to a wealthy and distinguished family. His father was secretary and adviser to three successive princes of Orange, and his elder brother, Constantine, succeeded to this responsible office and went to England in 1688 with William of Orange. Christian showed early signs of a very remarkable mind and his father, as was usual in those days, educated him as rapidly and widely as possible. He was instructed in music as well as in the literary subjects which were customary at the period, and at the age of thirteen he was introduced to the construction of machines, an occupation for which he gave evidence of extraordinary ability. At sixteen years he was sent to Leyden to study law, and twelve months later to the University of Breda, where he plunged into mathematics, which always remained his favorite study. His first mathematical essays brought him to the notice of Descartes. On leaving the university, in 1649, he traveled with the Count of Nassau, and on his return to Holland set to work with enthusiasm on the most abstruse mathematical problems and on the mechanical inventions which are his chief title to fame. He published treatises on the quantities of the parabola and other geometrical figures, and at the same time began to grind and polish lenses for large telescopes, becoming the maker of instruments superior to any which existed at the time. It was with a twelve-foot telescope of his own handiwork that he not only discovered, in 1656, a sixth satellite of Saturn, Titan, but determined the period of its revolution. Huygens was the first to apply the pendulum to clocks, and to employ the device to determine the acceleration of gravity. It would be impossible even to enumerate the many astronomical discoveries which Huygens initiated, but certainly his most illustrious observations were those by which he first understood the nature of Saturn's rings. That planet had been a most baffling puzzle to astronomers ever since the telescope had revealed its apparently triple form, for Saturn changes remarkably in appearance according as the rings are viewed edgewise—when they disappear, except to the most powerful telescopes—or are seen more or less in face. Having by far the best optical instrument of his age, Huygens was able, in 1656, to announce the true relation of Saturn and its rings, though he was not able to separate the ring-system into its component members. In "The System of Saturn", wherein the whole matter, so far as at that time disclosed, was clearly set forth, he records other important astronomical observations. For example, he describes the bands on Jupiter, the principal markings of Mars and the Orion nebula. He was able also to define the very significant fact that, even in the largest telescopes, the stars, as distinguished from planets, have no diameter, but are mere points of light without dimensions. From this discovery he very naturally turned to the determination of the diameters of the several planets, and invented an instrument for the purpose. This, as modified later, became the micrometer which is used in modern observatories. It should be said, however, that a previous worker in the same field, Gascoigne, who was killed at the battle of Marston Moor in 1644, had constructed a true filar micrometer, but the invention remained unknown for many years afterward. Huygens visited England in 1660, and was admitted a member of the Royal Society. It was a time of general scientific revival, and the French Academy of Sciences, which was being founded about the same time, persuaded Huygens to take part in its direction, offering him at the same time princely remuneration. He acceded to their request, and remained in Paris until the year 1681, when he resigned his position on account of failing health, and returned to his native Netherlands. To a man of this type, Sir Isaac Newton's *"Principia"*, which he secured in 1689, was a revelation and a delight. His first impulse was to cross the Channel once more, and get to know the author and, having done so, he wrote and published two important works, under the influence of Newton, on light and on weight respectively. But he was now coming near to the end of his days. His last illness began early in 1695, and he died on July 8 of the same year. The work he did was much more copious than can be even suggested by this brief notice. He did important work in optics and he is regarded as the founder of the waves theory of light. He had few equals in higher mathematics, and Newton praised his style and methods of mathematical exposition, and paid him the compliment of confessed imitation. He was a great expounder of the differential calculus, initiated by Pascal but still at this time generally unknown. Another important work dealt with the movements resulting from percussion, being an inquiry closely related to the modern science of ballistics. A lighter and more fanciful late work, entitled *"Consmotheoros"*, consists of conjectures on the physical constitution of distant worlds and their inhabitants. Using Galileo's discovery of the isochronicity of the pendulum, Huygens invented a serviceable clock powered by weights, a truly remarkable invention for which the world had been waiting through centuries of badly measured time. His speculations contributed concretely to the eventual development of the law of the conservation of energy as perfected by Helmholtz and others some 150 years later. He also, with brilliant insight, propounded a wave theory of light, which, however, was in conflict with Newton's corpuscular theory of light and, consequently, was brushed aside for the time. Although Huygens could be called a Cartesian, he was unable to accept Descartes's Universe, reasoning that particles of matter move in a vacuum. Huygens believed particles to be indivisible and completely elastic in collisions. Like the Greek Democritus, he suspected all objects were formed by various arrangements of the several classes of particles he imagined to exist. His reliance on mechanistic processes caused him to oppose Newton's *Principia* strongly because he could not imagine attraction as a basic principle due to its nonmechanistic origins. Huygens believed that heat as well as the light it produces are forms of vibration. Yet he realized that heat, unlike light, does not travel in a straight line. Huygens' attempt to explain the particle-like properties of light, without resorting to Newton's corpuscular theory, led to his own wave theory of light. As a lens-maker, Huygens was familiar with the law that governs refraction, the bending of light when it passes from air to water or from one transparent medium to another. His demonstration that waves also travel in straight lines and obey the observed laws of reflection and refraction enabled him to show how the principal laws of optics follow from his wave theory as well as from Newton's particle theory. Huygens believed light was transmitted through an ether medium consisting of minute particles. Light did not actually consist of matter but was formed by the displacement of ether particles. What is now known as Huygens's principle concluded that while the displacement of a single ether particle could not be detected, the "waves" created by the overlapping disturbances of great numbers of these particle collisions were sufficient to create visible light.

Hypatia. (b. c. 370; Alexandria, Egypt; d. 415; Alexandria). Greek mathematician. Hypatia was the daughter of Theon of Alexandria, the author of a well-known commentary of Ptolemy. Although little is known about her, she appears to be, with the exception of the alchemist Marie the jewess, the only named woman scientist of the Greek world. In 400 she was reported to be head of the Neoplatonic school in Alexandria. To her have been attributed commentaries on Ptolemy's *Almagest,* Diophantus's *Arithmetic,* and Appolonius's *Conics,* none of which have survived. Learning and science came to a violent conclusion in Alexandria and in the West, as did Hypatia. In conflict with Cyril, bishop of Alexandria, through her friendship with Orestes, the Roman prefect of the city, she was seized by the Christian mob and savagely mutilated and killed.

I-Hsing. (b. *c.* 681; d. *c.* 727). Chinese mathematician and astronomer. I-Hsing was a Buddhist monk around whom many legends have grown. Only a small portion of his work has survived so it is difficult to appreciate it in detail. There is, however, no reason to doubt his involvement in two major astronomical achievements. In the period 723-26, in collaboration with the Astronomer

Royal, Nankung Yueh, expeditions were organized to measure, astronomically, the length of a meridional line. Over a distance of 1553 miles (2500 km) along this line, simultaneous measurements of the Sun's solstitial shadow were made at nine stations. The estimated length of a degree, on the basis of their measurements, was far too large and it must be supposed that some systematic error in the method of observation was taking place. However, when it is appreciated that research expeditions to determine the length of a meridional degree were not organized in Europe until the 17th century, the amazing nature of I-Hsing's work can be appreciated. He also probably anticipated Su Sung in the use of an escapement in an astronomical clock. It was described in a 13th-century encyclopedia: "Water, flowing into scoops, turned a wheel automatically, rotating it one complete revolution in one day and one night". This turned various rings representing the motion of the celestial bodies. It was soon reported to be corroded, relegated to a museum, and to have fallen into disuse.

J

Jacobi, Karl Gustav Jacob. (1804-51), German mathematician who, independently of Abel, made dramatic progress in the theory of Elliptic Functions, number theory (in which his work was admired by Gauss), differential determinants, and various branches of analysis, geometry, and mechanics. He was first an Extraordinary and then an Ordinary Professor at the University of Königsberg (1827-42), and later at Berlin, and died of smallpox in 1851. Jacobi was a student in Berlin and became a lecturer at Königsberg where he managed to attract the favorable attention of Karl Friedrich Gauss. He was a superb teacher and had an astonishing manipulative skill with formulae. He made a brief but disastrous foray into politics that resulted in his losing a pension he had been granted by the king of Prussia. Jacobi's most important contributions to mathematics were in the field of elliptic functions. Niels Hendrik Abel had partially anticipated some of Jacobi's work, but the two were equally important in the creation of this subject. Jacobi also worked on Abelian functions and discovered the hyperelliptic functions. He applied his work in elliptic functions to number theory. Jacobi worked in many other areas of mathematics as well as the theory of functions. He was a pioneer in the study of determinants and a certain type of determinant arising in connection with partial differential equations is known as the *Jacobian* in his honor. This work was the result of his interest in dynamics, in which field he continued and developed the work of William Hamilton, and produced results that are important in quantum mechanics.

James Clerk Maxwell. James Clerk Maxwell is a name that would surely be high on any physicist's list of the half-dozen or so of the greatest intellects in the history of sciences. He was born in Edinburgh in 1831 and died in 1879 of abdominal cancer in Cambridge, England. His father was trained as a lawyer but was practical-minded; while a member of the Royal Society of Edinburgh, he had published a paper proposing a printing press with an automatic feeding devices. At fourteen James did his first original research on ovals. He matriculated at Cambridge University at nineteen and was elected to the Select Essays Club, a twelve-man honor society, whose members included at various times Alfred Tennyson, Alfred North Whitehead and Bertrand Russell. He took his first job as a professor of mathematics in Aberdeen in 1856. Maxwell's first major discovery was a well-reasoned explanation of the rings of Saturn, which he showed to be composed of small particles capable of orbiting the planet without disintegrating. A few years later he and then, ill later, Ludwig Boltzmann devised the kinetic theory of gases that bears both their names. In 1871 he became the first professor of experimental physics at Cambridge, but contemporaries report that his lectures were far too difficult for his students and were usually poorly attended. While teaching at Cambridge he organized the famous Cavendish Laboratory and was its director for the rest of his life. What is known as the Maxwell-Boltzmann kinetic theory of gases depended on an equation Maxwell had formulated to show the distribution of velocities of atoms in a gas at a particular temperature. An increase in the temperature would increase the valocities of the atoms while a decrease in temperature would similarly reduce their velocities. Maxwell's "demon" was introduced to highlight at differences between the fluid theory of heat and own molecular theory of heat. The fluid theory was based on Clausius's second law of thermodynamics and held that heat would inevitable flow from a warm reservoir to a cold reservoir. Maxwell showed how his molecular theory would work by pretending that his demon operated a door separating two temperature reservoirs so as to permit the slower particles to accumulate in one reservoir and the faster particles to accumulate in the other reservoir. The former reservoir would become colder and the letter reservoir steadily warmer—violating Clausius's second law. Although Maxwell's demon was only a literary device, it showed how the movements of molecules could conceivably operate in such a way as to reverse the second law after a long period of time. Maxwell's scientific masterpiece was his development of his mathematical version of Faraday's experimental work in the electromagnetic field. He showed that at each point in the world there exists an electromagnetic field consisting of electric and magnetic oscillating vectors (quantities having direction). As a field is simply a region under the influence of a physical force, it can be pictorially represented by a set of curves known as lines of flux. The density of these lines at a given point represents the local strength of the field. The direction in which these lines are oriented indicates the direction of the force stemming from the field. What we call a vector field may be represented by an arrow; the point indicates its direction and the length gives its magnitude. A familiar example of a vector field is the wind, which may be represented at each point on a weather map by an arrow depicting both its direction and magnitude. In 1864 Maxwell derived a set of four partial differential equations that are usually referred to as "simple equations." These four equations describe the behavior of the electric and magnetic vectors at every point and show how the electric and magnetic fields are related to each other. Simple they are to people who know their calculus, but other must take on faith the mathematicians' assurances that these differential equations prove the impossibility of having an electric field a magnetic field alone. They invariably exist together, directed perpendicularly to each other. (A static field can exist by itself, as can a static magnetic field; a moving electric change generates a magnetic field and a moving magnetic pole generates an electric field.) Maxwell found, as his equations show, that a changing electric field produces a change in its companion magnetic field and vice versa, with the field expanding outward in all directions. He proved that this field propagates at the velocity of some 186,000 miles per second—the same velocity at which light travels. This discovery led to the inescapable conclusion (eventually proved by Hertz) that light itself is part of a vast electromagnetic spectrum. Maxwell, who died at the comparatively early age of forty-eight, left a magnificent memorial to himself: with all the revolutionary changes that have taken place in physics since his death in 1879, there is one brief page that is as valid today as the day it was written: Maxwell's equations. To understand light was a great puzzle to scientists right from the time of Newton. For this, different scientists gave different theories. Some described light as particles and others as waves. Most of the scientists thought of ether as the medium through which light travels. The theories which were used to explain light were incomplete in one way or the other. This problem was solved by a British physicist, James Clerk Maxwell. He developed the electromagnetic theory of light which stood solid as a rock on all experimental fronts. This theory alone made Maxwell one of the world's greatest scientists. Nobody thought a simple child born in a village would develop a theory of science which would not only bring him world fame, but also resolve many problems of physics. This great scientist was born in Edinburgh. His father was a rich man. When he was only eight his mother expired, and his father brought him up. Right from his childhood Maxwell was very much attached to nature. He felt thrilled at the sight of lakes, mountains, and springs. He would often get totally lost in the colorful scenes of nature. Nature played an important role in developing his talent. It might be a gift of nature that, later he succeeded in giving an equation for various colours. There is an idiom that coming events cast their shadows before, which became true for Maxwell. The signs of intelligence in him started appearing when he was a child. At the age of 15, he developed a mechanical method to make cartesian oval. Cartesian oval is a type of curvature in geometry. Maxwell wrote a paper on this method which was published by the Royal Society of Edinburgh. When he was 18, he did significant research related to the equilibrium between rolling curves and elastic solids. He presented these research papers also before the Royal Society. He wrote a sensational paper related to the rings of Saturn, for which he was bestowed the Adam Award. The actual development of Maxwell's talent took place when he came in contact with the

Scottish scientist, Nichol. Nichol proved to be a great teacher for him. Maxwell started doing research in Nichol prism which proved very useful in developing Maxwell's research capability. with Nichol, he published a research paper on colour blindness. The most famous discovery of Maxwell was the electromagnetic theory of light. For this work, he got the inspiration from Michael Faraday. Although Faraday had himself known that light is electromagnetic in nature, but being not educated he was unable to give any mathematical base for this. Maxwell conducted many experiments to understand the electromagnetic nature of light and later gave the mathematical form to this. He later published his findings related to this subject in a paper called 'Dynamical Theory of Electromagnetic Field'. After eight years, he published a book on the subject named 'Treatise on Electricity and Magnetism'. Maxwell's electromagnetic waves also propagated the elastic hypothetical medium ether. Later, the concept of other was totally rejected, but his Electromagnetic Theory is true even today. Apart from developing the Electromagnetic Theory, he made a disc which is known as Maxwell Disc. He also worked on Kinetic Theory of gases, and Maxwell Boltzmann Statistics. He worked in the world famous Cavendish Laboratory of Cambridge and was made the first Cavendish professor of Physics at Cambridge in 1871. He strongly believed that real science should aim at helping the common man. Apart from being a great scientist, Maxwell was also good at swimming, gymnastics and horse riding. He was a good too. This great scientist died in 1879 while serving the cause of science. Although Maxwell is not with us today, his electromagnetic theory will always serve the mankind.

Jeans, Sir James Hopwood. (b. Sept. 11, 1877; London; d.Sept. 16, 1946; Dorking, England). British mathematician, physicist, and astronomer. Jeans, the son of a journalist, graduated from Cambridge University in 1900 and obtained his MA in 1903. After lecturing at Cambridge, 1904-05, he became professor of applied mathematics at Princeton University, 1905-09, and, back in England, Stokes Lecturer in applied mathematics at Cambridge (1910-12). There followed a period of writing and research during which his interest turned to astronomy. In 1923 he became a research assistant at the Mount Wilson Observatory, Pasadena, California, where he worked until 1944. Jeans was also professor of astronomy at the Royal Institution in London from 1935 until his death. He was knighted in 1928. Jeans is best known as an astronomer and as a writer both of popular books on science and of several excellent textbooks. His earlier books were devoted to physics and included *Dynamical Theory of Gases* (1904) and *Mathematical Theory of Electricity and magnetism* (1908). His serious astronomical works included *Problems of Cosmogony and Stellar Dynamics* (1929) and *Astronomy and cosmogony* (1928) while among his popular books were *The Universe Around Us* (1929) and *The Mysterious Universe* (1930). Jeans pioneered various ideas in astronomy and astrophysics. He showed that Pierre Simon Laplace's theory of the origin of the solar system, in which the Sun and planets condensed from a contracting cloud of gas and dust, was untenable. In collaboration with Harold Jefferys he proposed a new view in its place. According to this 'tidal theory' a star had passed close by the newly formed Sun and the planets had been formed from the cigar-shaped filament of material drawn away from this star. Jeans and Jeffreys based their theory on a similar idea proposed earlier by Thomas Chamberlin and Forest Ray Moulton. The tidal theory was eventually superseded in the 1940s by revamped versions of Laplace's nebular theory. Jeans also investigated other astronomical phenomena, among them spiral nebulae, binary and multiple star systems, and the source of energy in stars, which he concluded involved redioactivity.

Jeffreys, Sir Harold. (b. Apr. 22, 1891; Britley England). British astronomer, geophysicist and mathematician. Jeffreys was educated in Newcastle upon Tyne and at Cambridge University. After graduating in 1913 he was made a fellow of his college. He was reader in geophysics (1931-46) before elected to the Plumian Professorship of Astronomy and Experimental Philosophy where he remained until his retirement in 1958. In 1924 Jeffreys produced one of the fundamental works in geophysics of the first half of the 20th century. *The Earth: Its Origin, History, and Physical Constitution.* In this he argued forcibly against Alfred Wegener's proposed theory of continental drift. He demonstrated that the forces proposed by Wegener were inadequate. This did much to inhibit interest and research into drift theory for a while but much new evidence in its favor has since been uncovered. Jeffreys was also joint author, with Keith Bullen, of the *Seismological Tables* (1935). These, more frequently known as the JB *Tables,* were revised in 1940 and are the present standard tables of travel times of earthquake waves. They allow observers to determine from the elapsed time between the arrival of the primary (P) waves and the secondary (S) waves the distance between the observer and the earthquake. Jeffreys' work in astronomy included studies on the origins of the Universe. He developed James Jeans's theory of tidal evolution. He also devised models for the planetary structure of Jupiter, Saturn, Uranus, and Neptune. He was knighted in 1953.

Jerome Cardan. Jerome Cardan who is often referred to as Hieronymus Cardanus or Girolamo Cardano, Italian mathematician, astrologer, philosopher, gambler and charlatan, was born in 1500, the illegitimate son of Facio Cardan. His life was an uninterrupted course of misery, suffering and struggle. Beaten by his father till he was too weak to endure any more, in his youth he conceived the idea of becoming a Franciscan monk and even commenced his novitiate. Then, perhaps coerced by his father, a doctor, he studied medicine in Pavia, Padua and Venice. The young man married when he was thirty-two a poor girl, as wretched as himself, Nevertheless for a time he conquered had fortune and successively taught mathematics in Milan in 1534, and medicine in Pavia in 1540. Before dying in Rome in 1576, he had the grief of seeing his son die on the scaffold for having poisoned his wife. From his garrulous disquisition upon himself we are presented with an almost unbelievable mixture of hideous vice and eminent virtues. "Nature created me deft in the work of my hands, the philosophic spirit, apt in the study of science, with good taste and good character, voluptuous, gay, pious, inconstant, the friend of wisdom, prone to meditation, inventive, full of courage, enemy of authority, and of moderation, curious in all relating to medicine, eager for the miraculous, ... living from day to day, never forgetting an injury, envious, sad, treacherous, ... magician, enchanter, ... solitary, unsympathetic, austere, ... obscene, lying, obsequious." Beyond this enumeration he boasts in other parts of his works of his possessing hyperphysical qualities, voluntarily he enters into a trance. Dreams warn him of all that will happen. Finally, the future reveals itself to him by the marks that appear on his nails. Like Faust and like Socrates he possesses a familiar spirit whom he calls "Tetim". In Cardan's lifetime the spirit of inquiry that begot the Reformation was finding its way into the until then barren fields of many sciences. "Though he worked for the future he was not before his time. It was said after his death that no other man of his day could have left behind him works showing an intimate acquaintance with so many subjects. He sounded new depths in a great many sciences, brought with into the dullest themes, dashed wonderful episodes into abstruse treatises upon arithmetic. Jerome, however had not a whole mind, and the sick part of him mingled its promptings with the sound in all his writings." The *"Ars Magna"* (1545), by far his greatest work, contains the celebrated solution of the cubic equation. All his writings are difficult to analyze, for they are lengthy, tautological and lacking in arrangement; he wrote a treatise *"De rerum varietate,"* and *"De subtilitate rerum"*. His medical works occupy more than 4 volumes in folio of his *"Opera omina,"* and consist of a number of short treatises unrelated to one another. His last book, written at Rome, is upon himself and was the summing up of his intellectual accounts with the world. "There is one short chapter of rapid narrative and all the rest is self-dissection; it contains a chapter on his vices and another on his virtues, one on his honors, one on his disgraces, a long one on his friends, a very short one on his enemies." Cardan died at Rome on the 20th of September, 1576, when he was 75 years old, and his body was deposited in the Church of St. Andrew. Afterwards, probably by his grandson, it was removed to Milan, to be buried at St. Marks.

Johannes Kepler. Often regarded as the founder of modern astronomy, Johannes Kepler accomplished the very difficult talks of purging the increasingly unhelpful Greek astronomical legacy from sixteenth century science and improving the mathematical precision of the Copernican system by dropping circular planetary orbits in favor of ellipses. Kepler's principal scientific discoveries were his three planetary laws, discussed in the introduction to this section. His laws marked a crucial turning point in the evolution of Western science because they were the first modern "natural" laws; each law was derived numerically from Tycho Brahe's observations and then expressed mathematically so that it could be verified. They orpedoed the Aristotelian doctrine of circular motion that had stifled cosmology for nearly two millennia and swept away the last vestiges of the Ptolemaic epicycles. Like Galileo, Kepler wanted to replace the mystical orientation of medieval cosmology with a more mechanistic conception of nature. In many ways Kepler laid the foundation for the triumphant emergence of classical mechanics in Newton's theories of absolute space and time and universal gravitation. One of Kepler's more curious obsessions throughout his life was the presumed relationship between the five known planets of the solar system and the five Platonic solids—the tetrahedron or triangular pyramid, the cube, the octahedron, the dodecahedron, and the icosahedron. Since each form is perfectly symmetrical, it can be inscribed in or superscribed about a sphere. Kepler spent a great part of his life on what he thought was a discovery of prime—even divinely inspired—significance. Doubtless Kepler's interest in geometry compelled him to speculate about the importance of this relationship, although it contributed nothing to the development of physics. In addition to his mathematical talents, Kepler had an interest in astrology that contrasts curiously with his austere scientific personality. Kepler's posthumous work, *The Dream,* published by his son, is a strange work that partakes of fact and fancy and is unknown to most readers. Owen Gingerich writes that Kepler's *Dream* "is a curiously interesting tract ... [because] its fantasy framework of a voyage to the moon made it a pioneering and remarkably prescient piece of science fiction ... [and] its perceptive description of celestial motions as seen from the moon produced an ingenious polemic on behalf of the Copernican system."

John A. Wheeler and Remo Ruffini. John A. Wheeler and Remo Ruffini was born in Jacksonville, Florida, in 1911, the son of librarians. He obtained his Ph.D. in theoretical physics in 1933. After spending a year with Gregory Breit at NYU in New York and a year with Neils Bohr in Copenhagen, he came to the University of North Carolina where he remained until 1938. He then took a position at Princeton, where he worked as a professor of physics from 1947 to 1976, after which he moved to the University of Texas to take a similar academic position. Wheeler has made many contributions to physics; indeed, it is difficult to think of an area that has not been appreciably advanced by his theoretical work. While working with Bohr, Wheeler defined the mechanism of nuclear fission and predicted the fissibility of plutonium, the synthesis of which would later become so important to the American nuclear weapons program. Brief stints with the Chicago Metallurgical Laboratory and du Pont led to discoveries of how better to control a nuclear reactor; this work was to prove essential in the design of power plants for the fledgling nuclear power industry.

In 1937 Wheeler introduced the concepts of the scattering matrix and resonating group structure into nuclear physics. He continued his work in nuclear physics offering in 1953 with D. L. Hill a model of the atomic nucleus that distinguishes between the states occupied by the individual nucleons of the nucleus and the state of the nucleus as a whole. Wheeler joined Edward Teller and a number of other physicists at Los Alamos in 1951 to work on developing an explosive device fueled by heavy hydrogen. Over the next two years, the participants in the hydrogen bomb project, known as Project Matterhorn, worked out the nuclear physics and hydrodynamics of a variety of triggering devices. The practical results of this work were revealed in November 1952 with the explosion of the first hydrogen bomb on Eniwetok atoll. Much of the work done in Project Matterhorn and subsequently refined by those physicists, as well as others working at Los Alamos and Lawrence Livermore Laboratory, continues to guide the development of new generations of nuclear weapons. Wheeler turned his attention to general relativity theory in 1953, extending Einstein's view that particle are essentially geometrical knots in space. By assuming that the fields which formed these particles are real and the particles themselves are geometrical constructs. Wheeler carried this view to its logical extreme, arguing that in the context of quantum electrodynamics it makes no sense to talk about the interaction of particles since no particle can actually be specified. Wheeler argued that a close examination of the world line (an imaginary graph which plots the position of a particle against time) of an elementary particle would reveal a chaotic path of tiny zigzags that would describe in symbolic fashion the virtual paris of positive and negative electrons being created and annihilated all the time in the region of the particle. Since this process of creation and destruction occurs throughout space, Wheeler said that out concepts of particles and the interactions between particles are pointless ways to describe the dynamic character of nature. Instead the space-time continuum, that is "[m]atter, charge, electromagnetism, and other fields are only manifestations of the bending of space." In short, "physics is geometry." Wheeler's extrapolations of general relativity theory caused him to wonder about the physics of collapsed stars, particularly those stars with surface gravitational fields so intense that even light is inevitably drawn inward. He first coined the phrase "black hole" in a 1968 article for *The American Scholar* entitled "Our Universe: The Known and the Unknown." He has remarked that "black holes have no hair" because everything falling into a black hole loses its identity so that an outside observer cannot tell the nature of the object after it falls in. The present selection, coauthored with the physicist Remo Ruffini, surveys many of the properties of black holes. In some way it outlines Wheeler's own substantial contributions to this particular area of general relativity theory. General Relativity and the dynamics of gravitational collapse have been major interests of John Wheeler, perhaps the most versatile of modern American scientist and mathematician. His contributions to the study of black holes have already been noted. Wheeler has also tried to merge the so-called geometrodynamics of Einstein's theory of general relativity with quantum theory to show that space is not described by any single geometry such as the Euclidean geometry of parallel lines. Instead Wheeler suggests that space fluctuates among a variety of different geometries. According to Wheeler, there is a certain statistical probability that space-time will manifest itself in any one of a myriad of geometrical configurations. However, these fluctuations in space-time geometry are not visible as they are confined within the puny dimensions of the Planck length which is equal to 1.6×10^{-33} centimeter. This chaotic microscopic carpet of appearing and disappearing geometric bubbles of space-time has been referred to as the space-time foam. Wheeler argues that all the realities of the Universe are played out on a gigantic undulating stage of space-time foam known as superspace. It is a strange realm with in infinity of dimensions where each point fully describes a single Universe by designating the state of that Universe at a single instant in time, the coordinates of all the particles in that Universe, and the curvature of its space-time. This infinitude of parameters is Wheeler's infinite-dimensional superspace. By viewing general relativity and quantum theory as the twin organizers of the Universe, Wheeler had developed a "geometrodynamical interpretation of matter." As a result, a particle is considered to be a consequence of the fluctuations in the space-time foam taking place throughout superspace. Alternatively, a particle is simply a localized disturbance amidst the seething fluctuations and has no independent existence of its own. Indeed, the energy density of an elementary particle is estimated to be about 10^{18} times smaller than the energy density of the quantum fluctuations in the space-time foam. This vision is not one that appeals to atomists as it posits a continuum which is governed by fields. However, it may be viewed as a culmination of the shared vision of Faraday and Maxwell of a world in which even the supposedly indivisible atoms consist of nothing more than fields. This selection contains Wheeler's ideas about how the

Universe might be reprocessed were it to collapse upon itself. Like the geometries of the space-time foam, Wheeler argues that as all space and time hurtles toward an area of almost unthinkably small dimensions, there will be a "probability distribution of outcomes." The outcome will be shaped by the victor in the titanic battle to be waged between the imploding forces of gravity and the exploding spasms of energy on the Planck level, as well as a variety of other, less prominent parameters.

John James Waterston. The sad history of John James Waterston, the Scottish physicist, is unfortunately not unique in the annals of science. The sixth of nine children, Waterston was born in Edinburgh, Scotland, in 1811. His father was a manufacturer of stationery products who sent John James to the University of Edinburgh, where he studied mathematics and physics. At the age of twenty-one, he began working in London as a surveyor. Waterston then accepted a post as a hydrographer in the Admiralty. Waterston wrote this piece while serving as a teacher of East India Company cadets in Bombay. The work describes an independent discovery of certain results obtained by Daniel Bernoulli and supplies additional data of crucial importance to molecular research, recognizing that the mean kinetic energy of a molecule is a measure of the temperature of the gas and is the same for all molecules in a gas in equilibrium, no matter what their dimensions and mass. Although Hooke had associated constant "heat" with constant molecular speed, Waterston strengthened the connection between heat and speed by delcaring that temperature measures not the speed v but the product mv^2 which Leibniz called *vis viva*. Waterston's kinetic hypothesis began with Newton's optical theory that vibrating atoms emit light. By pointing to experiments suggesting that heat and light belong to the same family, Waterston concluded that vibrating atoms also emit heat. Waterston submitted this fine paper to the Royal Society for publication in 1845. Two referees flunked it and relegated it to the archives instead of returning it to its author, since that was the Society policy at the time. Had it been returned, Waterston might have had it published elsewhere. Rejection of the paper is all the more remarkable since the theory of heat was a vital issue of the day, with textbooks presenting both caloric and kinetic versions as possibly valid without their authors being able to choose definitely between them. Five years later the researches of Joule, Clausius and Maxwell had eliminated the caloric theory forever—a development that might have occurred earlier had Waterston's judges been less obtuse. The story has a belated happy ending: Over the years Waterston published a number of other papers, some of which led John William Strutt, Lord Rayleigh, vice-president of the Royal Society, to the old paper in the archives. He published it nearly fifty years after its composition, with a gracious and apologetic introduction which is included here.

Johannes Kepler. Often regarded as the founder of modern astronomy and mathematics Johannes Kepler accomplished the very difficult tasks of purging the increasingly unhelpful Greek astronomical legacy from sixteenth century science and improving the mathematical precision of the Copernican system by dropping circular planetary orbits in favor of ellipses. Kepler's principal scientific discoveries were his three planetary laws, discussed in the introduction to this section. His laws marked a crucial turning point in the evolution of Western science because they were the first modern "natural" laws; each law was derived numerically from Tycho Brahe's observations and then expressed mathematically so that it could be verified. They torpedoed the Aristotelian doctrine of circular motion that had stifled cosmology for nearly two millennia and swept away the last vestiges of the Ptolemaic epicycles. Like Galileo, Kepler wanted to replace the mystical orientation of medieval cosmology with a more mechanistic conception of nature. In many ways Kepler laid the foundation for the triumphant emergence of classical mechanics in Newton's theories of absolute space and time and universal gravitation. One of Kepler's more curious obsessions throughout his life was the presumed relationship between the five known planets of the solar system and the five Platonic solids—the tetrahedron or triangular pyramid, the cube, the octahedron, the dodecahedron, and the icosahedron. Since each form is perfectly symmetrical, it can be inscribed in or superscribed about a sphere. Kepler spent a great part of his life on what he thought was a discovery of prime—even divinely inspired—significance. Doubtless Kepler's interest in geometry compelled him to speculate about the importance of this relationship, although it contributed nothing to the development of mathematics and physics. In addition to his mathematical talents, Kepler had an interest in astrology that contrasts curiously with his austere scientific personality. Kepler's posthumous work, *The Dream,* published by his son, is a strange work that partakes of fact and fancy and is unknown to most readers. Owen Gingerich writers that Kepler's *Dream* "is a curiously interesting tract ... [because] its fantasy framework of a voyage to the moon made it a pioneering and remarkable prescient piece of science fiction ... [and] its perceptive description of celestial motions as seen from the moon produced an ingenious polemic on behalf of the Copernican system."

Joly, John. (b. Nov. 1, 1857; Hollywood, now in the Republic of Ireland; d. Dec. 8, 1933; Dublin). Irish mathematician and physicist. Joly was the son of a clergyman. He entered Trinity College, Dublin, in 1876 where he studied literature and engineering. He taught in the engineering school from 1883 and was appointed professor of geology and mineralogy in 1897, a post he held until his death. Joly's major geological work was in the field of geochronology. He first tried to estimate the age of the Earth by using Edmond Halley's method of measuring the degree of salinity of the oceans, and then by examining the radioactive decay in rocks. In 1898 he assigned an age of 80-90 million years to the Earth, later revising this figure to 100 million years. He published *Radioactivity and Geology* in 1909 in which he demonstrated that the rate of radioactive decay has been more or less constant through time. Joly also carried out important work in radium extraction (1914) and pioneered its use for the treatment of cancer. His inventions in physics included a constant-volume gas thermometer, a photometer, and a differential steam calorimeter for measuring the specific heat capacity of gases at constant volume.

John Napier. Many mathematical calculations are involved in the field of astrology, navigation, business, engineering and war sciences. John Napier devised an easier method for carrying our complex mathematical calculations. By Napier's method, complex operations of multiplication and division can be done easily and quickly. The method developed by John Napier is called Logarithm Method. By using logarithms, the complex multiplications are expressed as additions, and divisions as subtractions. The exponents on numbers are expressed as logarithm multiplications. After these conversions, the values are obtained from standard, logarithm tables and by addition and subtraction the result is obtained. Antilogarithm of this value provides the actual solution of the problem. Separate logarithm and antilogarithm tables were prepared by John Napier. The inventor of logarithms, John Napier, was a Scotish mathematician. He was born at Merchiston Castle, near Edinburgh, in 1550. At the time of his birth, his father was just 16. At the age of 13, Napier entered the University of St. Andrews, but his say appears to have been short, and he left the University without taking a degree. Very little is known about Napier's early life, but it is thought that he travelled abroad because, according to one custom, the sons of the Scottish landed gentry had to travel abroad. Napier was back home in 1571, and he stayed either at Merchiston or at girtness for the rest of his life. In 1572, Napier got married. His wife died in 1579. A few years after his wife's death, he married again. Napier's life was spent amid bitter religious diseases. In 1593, he wrote a book on the church of Rome, in which he wrote that Pope of church would destroy the world between 1688 and 1700. The 10 editions out of 21 printed so far were sold during his life time. Following the publication of this work, Napier occupied himself with the invention of secret instruments of war. He invented two kinds of burning mirrors. He also made a metal chariot from which shots could be discharged through small holes.

Napier devoted most of his spare time to the study of mathematics and science. The result of this was that his name became associated with mathematics. He also discovered an instrument which was used for addition, subtraction and calculating square roots. This instrument was known as Napier Rod. Although Napier discovered many things, greatest of these is logarithms. Napier started working on logarithms in 1593. He gradually elaborated his computational system whereby roots, products and quotients could be quickly determined. On this basis he developed logarithms table. Napier's invention of logarithms overshadows all his other mathematical work. He did many researches on trigonometry and Napier's Analogy. Napier died in Merchiston on April 4, 1617. The contributions of Napier to mathematics are contained in two treatises—Description of Marvelons Canon of Logarithms, which was published in 1614, and Construction of the Marvelons Canon of Logarithms, which was published in 1620 a few years after his death. Although Napier is not with us today, his method of logarithms is still widely used. In our country, students get acquainted with logarithms in secondary classes. This method is used by all countries of the world. Although computer had reduced the use of logarithms to some extent, the importance of this method can't be overlooked. It still remains the most economical method of speedy calculation.

Joliot-Curie, Irène. (b. Sept. 12, 1897; Paris; d. Mar. 17, 1956; Paris). French scientist. Irène Curie was the daughter of Pierre and Marie Curie, the discoverers of radium. She received little formal schooling, attending instead informal classes where she was taught physics by her mother, mathematics by Paul Langevin, and chemistry by Jean Baptiste Perrin. She later attended the Sorbonne although she first served as a radiologist at the front during World War I. In 1921 she began work at her mother's Radium Institute with which she maintained her connection for the rest of her life, becoming its director in 1946. She was also, from 1937, a professor at the Sorbonne. In 1926 Irène Curie married Frederic Joliot and took the name Joliot-Curie. As in so many other things she followed her mother in being awarded the Nobel Prize for distinguished work done in collaboration with her husband. Thus in 1935 the Joliot-Curies won the physics prize for their discovery in 1934 of artificial radioactivity. Irène later almost anticipated Otto Hahn's (q.v.) discovery of nuclear fission but like many other physicists at that time found it too difficult to accept the simple hypothesis that heavy elements like uranium could split into lighter elements when bombarded with neutrons. Instead she tried to find heavier elements produced by the decay of uranium. Like her mother, Irène Joliot-Curie produced a further generation of scientists. Her daughter, Hèléne, married the son of Marie Curie's old companion, Paul Langevin, and, together with her brother, Paul, became a distinguished physicist.

Jordan, (Marie-Ennemond) Camille. (b. Jan 5, 1838; Lyons, France; d. Jan. 20, 1922; Milan, Italy). French mathematician. Jordan studied in Paris at the Ecole Polytechnique where he trained as an engineer. Later he taught at both the Ecole Polytechnique and the Collège de France until his retirement in 1912. His interests lay chiefly in pure mathematics, although he made contributions to a wide range of mathematical subjects. Jordan's most important and enduring work was in group theory and analysis. He was especially interested in groups of permutations and grasped the intimate connection of this subject with questions about the solvability of polynomial equations. This basic insight was one of the fundamental achievements of the seminal work of Evariste Galois, and Jordan was the first mathematician to draw attention to Galois's work, which had until then been almost entirely ignored. Jordan played a major role in starting the systematic investigation of the areas of research opened up by Galois. He also introduced the idea of an *infinite* group. Jordan also passed on his interest in group theory to two of his most outstanding pupils, Felix Klein and Sophus Lie, both of whom were to develop the subject in novel and important ways.

Jordan, Ernst Pascual. (b. Oct. 18, 1902; Hannover, now in West Germany). German theoretical physicist and mathematician. Jordan was educated at the Hannover Institute of Technology and the University of Göttingen, where he obtained his doctorate in 1924. He left Göttingen in 1929 for the University of Rostock and after being appointed professor of physics there in 1935, later held chairs of theoretical physics at Berlin from 1944 to 1952 and at Hamburg from 1951 until his retirement in 1970. Jordan was one of the founders of the modern quantum theory. In 1925 he collaborated with Max Born (q.v.) and in 1926 with Werner Heisenberg (q.v.) in the formulation of quantum mechanics. He also did early work on quantum electrodynamics. He developed a new theory of gravitation at the same time as Carl Brans and Robert Dicke (q.v.).

Josiah Willard Gibbs. Josiah Willard Gibbs, who was born in 1839 in New Haven, Connecticut, passed virtually his entire professional career in New Haven and died there in 1903. His father was a noted philologist and professor of literature at Yale. Gibbs received his undergraduate education at Yale College, where he won a number of prizes in Latin and mathematics. In 1863 he became the first graduate student ever to receive a Ph.D. in engineering from Yale University. Following graduate study at Yale and three years as a tutor, he traveled to France and Germany, then the indispensable course for any serious American scientist. He came home in 1869, took a post as professor of mathematical physics at his alma matter (the first nine years of which were unpaid) and continued in that capacity until his death at the age of sixty-four. Throughout his life Gibbs rarely left New Haven except for summer holidays in northern New England and occasional scientific conferences. He spent his entire life in the house in which he had been born, which was located less than one block from the university. His fame rests on a rather small amount of published material—some 400 pages made up of papers on thermodynamics, all of them published in a three-year period in the *Transactions of the Connecticut Academy of Sciences.* His raw material was the thermodynamic principles developed by Carnot, Joule, Helmholtz and Kelvin; the originality of Gibbs's contribution was to apply them in a rigorously mathematical manner to chemical reaction. His major work, "On the Equilibrium of Heterogeneous Substances," extended the application of thermodynamics to chemical, elastic, electromagnetic and electrochemical systems. He was praised by Boltzmann as "one of the great American scientists, perhaps the greatest," for his contributions to the development of statistical mechanics and for the invention of the name of the new discipline as well. In the nineteenth century American physics was not taken very seriously by the rest of the scientific world. Not until the 1890s, when Friedrich Wilhelm Ostwald translated some of Gibbs's work into German and Henri Louis Le Châtelier translated it into French, were Gibbs's great contributions recognized throuthout much of Europe. Two years before his death he was awarded the Copley medal of the Royal Society and in 1950 he was elected to the Hall of Fame of Great Americans. Our selection, a preface to his book on thermodynamic, is one of Gibbs's very few nonmathematical writings.

K

Kendall, Maurice George. (b. Kettering, Northants., UK, 6.9.1907). British statistician, with experience in government, shipping, academia, commercial consultancy and the World Fertility Survey. Kendall is the great synthesizer and commentator in modern statistics. Postulating that the differing schools are all addressing themselves to different questions, he has made common a heteroclite form of statistician who follows different system in different situations. His *The Advanced Theory of Statistics* (3 vols, London, 1961-6; 4th ed. with A. Stuart, London, 1977-9) is the practitioner's Bible. He is also known for redeveloping rank correlation by inventing a new form (Tau) insensitive to the absence of a coherent underlying scale and extending Spearman's work to many observers and several variables (*Rank Correlation Methods*, London, 1948,[3] 1962).

Kepler, Johannes. In the evening of the 15th November, 1630, a small, slightly built man, who was not yet old, died in a small room in the house of a merchant in Ratisbon (Regensburg) in South Germany. The death of one of the greatest astronomers in history, of the man who had formulated the three fundamental laws of the movement of the planets, Johannes Kepler, took place unnoticed. Johannes Kepler was born in Weil in

Württemberg (South Germany) on 27th December 1571. He was the son of a poor innkeeper and in the natural course of events should have become a waiter in his father's business. But he was so unsuited for this kind of work that his parents decided to send him to study to be a Protestant pastor: this was a most fortunate decision for astronomy. Kepler, therefore, went to the famous seminary-university of Tübingen, where he studied theology. Here an event took place that was to determine his future: he met a professor who expounded the Copernican system to his. Nicholas Copernicus, a Polish/German scholar, had thirty years earlier published his very famous theory on the solar system, in which he rejected the theory of Claudius Ptolemy (150 A.D.), which put the immovable Earth in the centre of the universe with the Sun and planets rotating round it; Copernicus, on the other hand, maintained that the Sun and not the Earth is at the centre of the solar system, while the Earth is a planet like the others, all moving round the Sun. Kepler immediately perceived the correctness of this theory and became a convinced Copernican. His name soon became famous. So much so that in 1599 the celebrated astronomer Tycho Brahe invited him to come to Prague as his assistant. In 1600 Kepler settled in Prague; a few months later the great Brahe died and Kepler succeeded him as astronomer to Emperor Rudolph II. On clear nights he made observations of the planets with primitive optical instruments, then turned to his sheets covered with figures and studied and calculated for hours on end untiringly. Kepler proved that the helicentric system (from the Greek 'Helios', sun; i.e. which put the Sun in the centre) of Copernicus was the only one which correctly represented reality. By means of various complex calculations, he established the three great laws regarding the movement of the planets. These are: (1) The planets revolve round the Sun with a movement that is not circular but elliptical; the Sun is at one of the foci of the ellipses and is not in the centre of them. An ellipse is the figure obtained by cutting a cylinder with a slanting plane. (2) The velocity of a planet as it revolves round the Sun varies according to its distance from the Sun. When near, it travels more quickly, and as it moves further away it slows down. The planet shown in the illustration covers the two distances A and B in the same time so that to cover the distance A it will travel more quickly. As a result of this law the areas of the two triangles shown in the figure are equal to each other. (3) This law, which is the most difficult of the three, is best explained by an example: Mercury takes 88 days and the Earth 365 days to revolve once round the Sun. By multiplying each of the two figures by itself (i.e. by squaring them) we get the figures 7,744 and 133, 225. The second number is about seventeen times the first, i.e. their ration is 1 to 17. Let us ncw see a ratio of their distances from the Sun. Mercury is on the average 36 million miles away and the Earth 93 million. By multiplying these figures by themselves twice (i.e. by cubing them) we get the figures 46,656 and 814,357. The ratio between these two figures is still very nearly 1 to 17. Here then is the law: the squares of the periods of revolution of any two planets are in the same relation to each other as the cubes of their average distances from the Sun. These three laws are still fundamental in astronomy and represented a big step forward in human knowledge. Kepler was also concerned with physics and the earth's magnetism. He was also the first person to calculate latitude and longitude accurately. After a hard and embittered life, he died a lonely man. But now we know that he was a genius and, like Copernicus, Galileo and Newton, he showed man the great harmony that exists in nature. Kepler's grandfather had been the local burgomaster but his father seems to have been a humble soldier away on military service for most of Kepler's early youth. His mother was described by Kepler as "quarrelsome, of a bad disposition." She was later to be accused of witchcraft. Originally intended for the Church, he graduated from the University of Tübingen in 1591 and went on to study in the theological faculty. In 1594 he was offered a teaching post in mathematics in the seminary at Gratz in Styria. It was from his teacher, Mästlin, who was one of the earliest scholars fully to comprehend and accept the work of Nicolaus Copernicus, that the young Kepler acquired his early Copernicanism. In addition to his teaching at Gratz and such usual duties as mathematicians were expected to do in those days Kepler published his first book—*Mysterium conmographicum* (1596). The book expresses very clearly the belief in a mathematical harmony underlying the universe, a harmony he was to spend the rest of his life searching for. In this work he tried to show that the universe was structured on the model of Plato's five regular solids. Although the work verges on the cranky and obsessive it shows that Kepler was already searching for some more general mathematical relationship then could be found in Copernicus. He married in 1597 shortly before he was forced to leave Gratz when, in 1598, all Lutheran teachers and preachers were ordered to leave the city immediately. Fortunately for Kepler, he had an invitation to work with Tycho Brahe who had recently become the Imperial Mathematician in Prague. Tycho was the greatest observational astronomer of the century and he had with him the results of his last 20 years' observations. Kepler joined him in 1600 and although their relationship was not an easy one it was certainly profitable. Tycho assigned him the task of working out the orbit of Mars. Somewhat rashly Kepler boasted he would solve it in a week—it work him eight years of unremitting effort. Not only did Kepler lack the computing assistance now taken for granted but he was also working before the invention of logarithms. It was during this period that he discovered his first two laws, and thus, with Galileo, began to offer an alternative physics to that of Aristotle. The first law asserts that planets describe elliptic orbits with the Sun at one focus while the second law asserts that the line joining the Sun to a planet sweeps out equal areas in equal times. The laws were published in his magnum opus *Astronomia nova* (1610; New Astronomy). Tycho had died in 1601 leaving Kepler with his post, his observations, and a strong obligation to complete and publish his tables under the patronage of their master, the emperor Rudolph II. This obligation was to prove even more onerous and time consuming than the orbit of Mars. It involved dealing with Tycho's predatory kin, attempting, vainly, to extract money from the emperor to pay for the work, which he ended up financing himself, and trying to find a suitable printer. All this, it must be realized, was done against the background of the Thirty Years' War, marauding soldiery, and numerous epidemics. The *Tabulae Rudolphinae* was not completed until 1627 but remained the standard work for the best part of a century. While serving the emperor in Prague, Kepler had also produced a major work *Optics* (1604), which included a good approximation of Snell's law, improved refraction tables, and discussion of the pinhole camera. In the same year he observed only the second new star visible to the naked eye since antiquity. He showed, as Tycho had done with the new star of 1572, that it exhibited no parallax and must therefore be situated far beyond the solar system. He studied and wrote upon the bright comet of 1607—later to be called Halley's comet—and those of 1618 in his *Three Tracts on Comets* (1619). His final work in Prague, the *Dioptrics* (1611), has been called the first work of geometrical optics. In 1611 Kepler's wife and son died, civil war broke out in Prague, and Rudolph was forced to abdicate. Kepler moved to Linz in the following year to take up a post as a mathematics teacher and surveyor. Here he stayed for 14 years. He married again in 1613. While in Linz he produced a work that, starting from the simple problem of measuring the volume of his wine cask, moved on to more general problems of mensuration—*Nova stereometria* (1615). One further crisis he had to face was his mother's trial for witchcraft in Würtemburg. The trial dragged on for three years before she was finally freed. His greatest work of this period *Harmonices mundi* (1619) returns to the search for the underlying mathematical harmony expressed in his first work of 1596. It is here that he stated his third law: the squares of the periods of any two planets are proportional to the cubes of their mean distance from the Sun. After the completion of the Rudolphine tables Kepler took service under a new patron, the Imperial General Wallenstein. He settled at Sagan in Silesia. In return for the horoscopes Wallenstein expected from him, Kepler was provided with a press, a generous salary, and the peace to publish his ephemerides and to prepare his work of science fiction—*A Dream, or Astronomy of the Moon* (1634). He left Sagan in 1630, during one of Wallenstein's temporary military setbacks, to see the emperor in Ratisbon hoping for a payment of the 12 000 florins still owed him. He died there of a fever a few days later. As a scientist Kepler is of immense importance. Copernicus was in many ways a traditional thinker, still passionately committed to circles. Kepler broke away from this mode of thought and in so doing posed questions of planetary motion that it took a Newton to answer.

Klein, Christian Felix. (b. Düsseldorf, N. Rhine-Wesphalia, Germany, 25. 4. 1849; d. Göttingen, Lower Saxony, 22. 6. 1925). German mathematician and educator. After holding chairs at Erlangen, Munich and Leipzig, Klein was called to Göttingen in 1886 and passed the rest of his career there. In collaboration with F. Althoff, the Prussian minister of culture, he encouraged the postgraduate training of women; many men also took doctorates under his supervision. He was a leader in planning the collected works of the 19c mathematician C. F. Gauss; the *Encyklopädie der mathematischen Wissenschaften mit Einschluss ihrer Anwendungen* [Encyclopedia of mathematical knowledge] (1898-1935), a vast multi-authored project reporting on the state of mathematics of the time; and the International Commission of Mathematics Education, which produced a series of important monographs in the 1900s and 1910s. Klein's first major contribution to mathematics was his inaugural lecture at Erlangen in 1872, now known as the 'Erlangen programme', in which he proposed to classify the various geometries in terms of the properties of group theory. His work here was important not only for geometry but also for raising in mathematics the status of group theory itself. In concurrent studies he also produced so-called 'projective' models for Euclidean and non-Euclidean geometries. In the 1880s he applied the ideas of group theory to various aspects of differential equations and the theory of functions. Some of this work was produced in competition with J. H. Poincare; and the stress under which he was placed seems to have affected his creative powers. Thus, when he came to Göttingen his major mathematical achievements were behind him, and his energies turned more to the educational and public enterprises described above. His papers are collected in *Gesammelte mathematische Abhandlungen* [Collected mathematical papers] (3 vols, Berlin, 1921-3). He attempted to classify and unify geometry by means of a general group-theoretic definition (*Kleinian geometry*) and was influential in the study of Elliptic Functions. He was professor at Erlangen from 1872, then between 1880 and 1913 was Professor at Munich, Leipzig, and Göttingen, and played a leading role in the scientific and mathematical community of his time. His also wrote about mathematics for the general public, and founded a mathematical encyclopaedia, which he continued to supervise until his death.

Klein, (Christian) Felix. (b. Nov. 25, 1849; Düsseldorf, now in West Germany; d. Jan. 22, 1925; Göttingen, now in West Germany). German mathematician. Klein, one of the great formative influences on the development of modern geometry, studied at Bonn, Göttingen, and Berlin. He worked with Sophus Lie—a collaboration that was particularly fruitful for both of them and led to the theory of groups of geometrical transformations. This work was later to play a crucial role in Klein's own ideas on geometry. Klein took up the chair in mathematics at the University of Erlangen in 1872 and his inaugural lecture was the occasion of his formulation of his famous *Erlanger Programm*, a suggestion of a way in which the study of geometry could be both unified and generalized. Throughout the 19th century, with the work of such mathematicians as Karl Friedrich Gauss, Janos Bolyai, Nikolai Lobachevsky, and Bernhard Riemann, the idea of what a 'geometry' could be had been taken increasingly beyond the conception Euclid had of it and Klein's ideas helped show how these diverse geometries could all be seen as particular cases of one general concept. Klein's central idea was to think of a geometry as the theory of the invariants of a particular group of transformations. His *Erlanger Programm* was justly influential in guiding the further development of the subject. In particular Klein's ideas led to an even closer connection between geometry and algebra. Klein also worked on projective geometry, which he generalized beyond three dimensions, and on the wider application of group theory, for example, to the rotational symmetries of regular solids. His name is remembered in topology for the *Klein bottle*, a one-sided closed surface, not constructible in three-dimensional Euclidean space. In 1886 Klein took up a chair at Göttingen and was influential in building Göttingen up into a great center for mathematics.

Klingenstierna, Samuel. (b. 1698; Linköping, Sweden; d. Oct. 26, 1765; Stockholm). Swedish mathematician and physicist. Before embarking on his mathematical and scientific studies Klingenstierna studied law at Uppsala. He was appointed secretary to the Swedish treasury (1720) but also had interests in philosophy and science and was allowed to continue his studies at Uppsala. In 1727 he was awarded a scholarship, which enabled him to travel in Europe. He traveled to Marburg where he studied with the Leibnizian philosopher Christian Wolff and also to Basel to study mathematics with Johann I Bernoulli. Klingenstierna became professor of mathematics at Uppsala and later professor of physics there (1750). His last appointment was the highly prestigious one of tutor to the crown prince (1756-64). Klingenstierna's most notable scientific work was in the field of optics. He was able to show that some of Newton's views on the refraction of light were incorrect and made practical use of this discovery in producing designs for lenses free from chromatic and spherical aberration.

Kolmogorow, Andrey Nikolayevich. (b. Tambov, Russia, 25. 4. 1903). Russian mathematician. Professor at Moscow State University since 1931. Kolmogorov has been immensely influential in all branches of modern probability theory. He is also known for his work in topology, in ring theory, and in functional analysis, where he first came to notice by demonstrating an integrable function having a divergent Fourier series. His most fundamental work in probability theory was his restatement of its foundations as a series of axioms using the language of measure theory; the probability of a random variable falling in a set is expressed as an integral over that set. Not only did this mean that many results in measure theory could be reinterpreted as theorems of probability theory, but more importantly, it freed the handling of probabilities from questions of interpretation. Since measure theory lies inside pure mathematics, its results apply indifferently in any system in which the axioms of probability in any sense hold (see *Osnovnye Ponyatiya Teorü Veroyatnostey* Moscow 1936; *Foundations of the Theory of Probability*. London. 1950). Kolmogorov developed the theory of stochastic processes (whose state at a given stage is a random variable whose distribution depends on knowable factors such as how long it has been going, or some of the past history of the system). Markov had developed a very useful class of processes in which the dependence on past history was minimal; Kolmogorov gave an extremely practical means of finding their distribution functions (his forward and backward equations) by looking at the probabilities of change in very small time intervals to obtain a set of differential equations. These have now been surpassed in elegance, but not at all for practical everyday use in a multitude of situations. He also obtained more general results, such as the law of the iterated logarithm. Related work of his in generalized mechanics relates to the ergodic states of systems; those which are not only possible but always likely to happen. Kolmogorov also worked on the theory of finite automata and produced related work on the theory of randomness. A sequence is taken to be random if the smallest program that will generate it is as long as the series itself. This is not altogether satisfactory, but it seems that any definitive theory of randomness will have to embody Kolmogorov's work.

Kolmogorov, Andrel Nikolalevich. (b. Apr. 25, 1903; Tambov, now in the Soviet Union). Soviet mathematician. Kolmogorov was educated at Moscow State University, graduating in 1925. He became a research associated at the university, later a professor (1931), and in 1933 was appointed director of the Institute of Mathematics there. He has made distinguished contributions to a wide variety of mathematical topics. He is best known for his work on the theoretical foundations of probability, but has also made lasting contributions to such diverse subjects as Fourier analysis, automata theory, and intuitionism. In 1933 Kolmogorov published his major treatise on probability, translated into English in 1950 as *The Foundations of the Theory of Probability.* The book is a landmark in the development of the theory, for in it he presents the first fully axiomatic treatment of the subject. It also contains the first full realization of the basic and underivable nature of the so-called 'additivity assumption about probability, first put forward by Jakob I Bernoulli. This claims simply that if an event can be realized in any one of an infinite number of mutually exclusive ways, the probability of the event is simply the sum of the probabilities of each of these ways. This assumption is fundamental to the whole measure-theoretic study of probability. Kolmogorov's interest in Luitzen Brouwer's intuitionism led him to prove that intuitionistic arithmetic, as formalized by Brouwer's disciple Arend Heyting, is consistent if and only if classical arithmetic is. In 1936 Kolmogorov settled a key problem in Fourier analysis when he constructed a function that is (Lebesgue) integrable, but whose Fourier series diverges at every point. In 1939, the same year in which he was elected an academician of the Soviet Academy of Sciences, he published a paper on the extrapolation of time series. This was later taken much further by Norbert Wiener and became known as 'single-series prediction'.

Kovalevskaya, Sofya Vasilyevna. (Sonya Kovalevski) (b, Jan. 15, 1850; Moscow; d. Feb. 10, 1891; Stockolm). Russian mathematician. Kovalevskay came from a cultured aristocratic background and studied mathematics privately. In 1869 she went to study at the University of Heidelberg, where her teachers included Gustav Kirchhoff and Hermann von Helmholtz. In 1871 she moved on to the University of Berlin to study with Karl Weierstrass. Since women were not allowed to attend lectures, Kovalevskaya studied privately with Weierstrass and a lifelong friendship developed between them. In 1874 when she took her doctorate (on partial differential equations) at the University of Göttingen, she had already produced original pieces of research on such varied topics as the constitution of the rings of Saturn and Abelian integrals, and had made important contributions to the theory of differential equations. However, in spite of her obvious outstanding talents and the determined backing of her influential friend Weierstrass, Kovalevskaya was unable to find an academic post anywhere in Europe, so strong was the prejudice against academic women at that time. She was thus compelled to return to Russia and settle down to family life. However after her husband's death the family's fortunes declined rapidly, and an academic job again became more of a necessity. This time Weierstrass was able to persuade the Swedish mathematician Mittag Leffer to offer Kovalevskaya a lectureship at Stockholm University. Her initial researches were in analysis, under Weierstrass's influence, and she wrote a memoir *on the Rotation of a Solid Body About a Fixed Point* (1988), which won her the coveted Prix Borodin from the

French Academy. Later she won a further prize from the Swedish Academy with another memoir on the same subject. Kovalevskaya had a considerable literary talent in addition to her mathematical gifts, and wrote numerous novels together with some autobiographical sketches.

Kronecker, Leopold. (b. Dec. 7, 1823; Liegnitz, now Legnica in Poland; d. Dec. 29. 1891; Berlin). German mathematician. Kronecker studied mathematics at Berlin but he did not become a professional mathematician until relatively late in life. He worked, highly successfully, as a businessman until he had made enough money to abandon commerce and devote himself fully to mathematics. He taught at Berlin from 1861, and, in 1883, was appointed professor. Outside mathematics Kronecker's interests were wide. He was a highly cultured man who used his wealth to patronize the arts. He also had a deep interest in philosophy and Christian theology, although he was not converted to Christianity until shortly before his death. Kronecker's mathematical work was almost entirely in the fields of number theory and higher algebra, although he also made some contributions to the theory of elliptic functions. His work on algebraic numbers was inspired by his constructivist outlook, which involved a distrust of nonconstructive proofs in mathematics and a suspicion of the infinite and all kinds of number other than the natural numbers. This attitude led him to rewrite large areas of algebraic number theory in order to avoid reference to such suspect entities as imaginary or irrational numbers. Kronecker's constructivism is summed up in a famous remark he made during an after-dinner speech: "God made the integers, all else is the work of man." His suspicion of nonconstructive methods led Kronecker into fierce controversy with two of the leading mathematicians of his day, Karl Weierstrass and Georg Cantor. His outlook anticipates to a considerable extent the views of the Dutch mathematician, L.E.J. Brouwer. Kronecker was also one of the first to understand thoroughly and use Evariste Galois's work in the theory of equations;. The *Kronecker delta function* is named for him.

L

Lagrange, Joseph Louis, Comte de. (1736-1813) Italian-born French mathematician and physicist who, in 1755, became Professor of Geometry at the Royal Artillery School at Turin, where he founded the Academy of Sciences. He later succeeded Euler as Director of Mathematics at the Berlin Academy of Science, returning to France after the death of Frederick the Great. His treatment of Isoperimetric Problems laid a foundation for the calculus of variations, and he made significant contributions to many other branches of mathematics, including probability theory, number theory, the theory of equations, and the foundations of group theory. Lagrange was born in Italy of French ancestry. At school he was at first more interested in the classics until he read an essay by the astronomer Edmond Halley on the calculus, which converted him to mathematics. By the age of 18 he was teaching at the Artillery School in Turin, eventually becoming professor of mathematics there. He also began a discussion group, which evolved to form the Turin Academy of Sciences in 1758. By 1760 Lagrange's reputation as one of Europe's greatest mathematicians was established; he received a prize from the Paris Academy of Science in 1764 for an essay on the liberation of the Moon. On the invitation of Frederick the Great he succeeded Leonhard Euler as mathematical director of the Berlin Academy in 1766, serving there until 1787 when, following Frederick's death, he moved to Paris and a post in the Academy of Sciences. Lagrange produced his greatest work, *Mècanique analytique* (1788; Analytical Mechanics), in Paris. This summarized the research in mechanics since Isaac Newton, based on Lagrange's own calculus of variations, and finally placed the mechanical theory of solids an fluids on a reigorous and analytical foundation. The book broke away from Euclidean tradition, and Lagrange commented in it that "one cannot find any diagrams in this work." Lagrange also made many contributions to astronomy and number theory. His work on the theory of equations helped Niels Abel in his later development of group theory. In astronomy he found a special solution to the three-body problem showing that asteroids will tend to oscillate around a certain point now called the *Largrangian point.* Much of the credit for the revolutionary introduction of the metric system is also due to him. Despite Lagrange's being technically a foreigner and also a friend of aristocrats he was held with respect during the French Revolution, which broke out in 1789, and was made president of the commission for metrication in 1793. He founded the mathematics department in the Ecole Normale in 1795 and later in the Ecole Polytechnique in 1797. Napoleon honored him, making him a count and a senator.

Lamb, Sir Horace. (b. nov. 29, 1849, Stockport, England; d. Dec. 4, 1934; Cambridge, England). British mathematician and geophysicist. Lamb studied at Cambridge University. His earliest interest had been in classics, but soon after arriving in Cambridge he discovered his true vocation lay in mathematics. Among his teachers were George

Stokes and the great mathematical physicist James Clerk Maxwell. Lamb's own interest in mathematics was very much shaped by contact with the ideas of these two men, and was almost entirely in applied mathematics. In 1875 he went to Australia to become the first professor of mathematics at the newly founded University of Adelaide. In 1885 he returned to England to take up the chair of mathematics at Manchester University, and held this post until his retirement in 1920. He was knighted in 1931. Among the great variety of fields of applied mathematics to which Lamb made important contributions are electricity and magnetism, elasticity, acoustics, vibrations and wave motions, statics and dynamics seismology, and the theory of tides and terrestrial magnetism. However it is for his work in fluid mechanics that Lamb is most celebrated. He wrote a book on the subject, *Hydrodynamics* (1895), which immediately became a classic and by 1932 had gone through six editions. Lamb's work in geophysics was also important. He wrote a paper in 1904 on the propagation of waves over the surface of an elastic solid, in which he virtually laid the whole theoretical foundations for modern mathematical seismology.

Lambert, Johann Heinrich. (b. Aug. 26, 1728; Mulhouse, now in France; d. Sept. 23, 1777; Berlin). German mathematician, physicist, astronomer, and philosopher. Lambert was the son of the poor tailor and was largely self-educated. He spent the early part of his life in various occupations, including teaching and bookkeeping, and followed his scientific interests in his spare time. He moved to Berlin in 1764 where he attracted the notice of Frederick the Great and became a member of the Berlin Academy. Lambert contributed to numerous branches of science and learning generally. His main mathematical achievement was to prove that 'π' is irrational. He did this by use of continued fractions and published the proof in 1768. He also sutdied the hyperbolic functions and introduced their use into trigonometry. Lambert did some remarkable work in non-Euclidean geometry, but this ramained totally unknown until the end of the 19th century when it was published. In addition to mathematics Lambert was an astronomer of note. The suggestion that there might be further galaxies beyond our own was first made by him and this was subsequently confirmed observationally by William Herschel. Lambert was the first to invent an accurate way of measuring light intensities and the lambert, a measurement of light intensity, was named for him. In his philosophical ideas Lambert largely developed the ideas of the great German rationalist philosopher Gottfried Leibniz and his chief philosophical work *Neues Organon* (New Organon), was published in 1764.

Laplace, Pierre Simon, Marquis de. (1749-1827), French analyst, probability theorist, and physicist, who was originally educated by the charity of neighbours, and is often regarded as the greatest exponent of celestial mechanics since Newton, providing that planetary perturbations do not disturb but maintain the stability of the solar system. He also proved that respiration is a form of combustion. His standing survived the fall of successive regimes: he was for six weeks Napoleon's Minister of the Interior, later became Chancellor of the Senate, was ennobled both under the Empire and by Louis XVIII, and was elected President of the *Académie Francaise*.

Lebesgue, Henri Léon. (b. Beauvais, Oise, France, 28.6.1875;

d. Paris, 26.7.1941). French mathematician. Lebesgue studied at the Ecole Normale Supérieure, where he came under the influence of Borel. This influence continued during his professional career, which he passed at the Collège de France. Taking up Borel's notion of measure, he developed in the 1900s a theory of the integral which was more general in conception and powerful in application than the previous theories (which are associated with the 19c mathematicians Cauchy and Riemann). His relations with Borel suffered later after some rather trivial priority disputes. Lebesgue applied his new theory to various problems in mathematical analysis, including the relationship between differentiation and integration (the fundamental relationship in the calculus), various properties of trigonometric series (including their relation to 'Fourier series') and the class of functions which could be defined by analytic expressions. These achievements are best seen as a major contribution to a whole generation of French mathematicians including not only Borel but also Denjoy, Frechet and Hadamard. In later years Lebesgue transferred some of his mathematical interests to other topics, although he did not attain the same level of importance. For example, while his ideas on topology were ingenious, they were outclassed by the achievements of Brouwer. During this period he also wrote some useful articles on the history and teaching of mathematics. His writings are collected in *Oeuvres scientifiques* (Scientific works] (5 vols. Geneva, 1972-3). He generalized the Riemann Integral to the Lebesgue integral by incorporating the concepts of Jordan Measure and Borel Measure in the of Lebesgue Measure. He established fundamental theorems of Lebesgue measure theory, applied Lebesgue Integration to the study of Fourier

Series, and worked on the development of abstract measure theory. He was elected to the French Academy of Sciences and the Royal Society of London.

Lebesgue Henri Léon. (b. June 28, 1875; Beauvais, France; d. July 26, 1941; Paris). French mathematician. Lebesgue studied at the Ecole Normale Supérieure. He obtained posts at Rennes (1902) and Poitiers (1906) universities, at the Sorbonne (1910), and at the Collège de France (1921). Lebesgue's extremely important contributions to measure theory and the theory of integration were stimulated by the earlier work of Emile Borel and by Camille Jordan's famous Cours d'Analyse. The importance of Lebesgue's work resides in the fact that he was the first mathematician to develop integration in a measure-theoretic context. This allowed natural generalizations of both concepts to be made. Lebesgue's definition of integral is considered, in many respects, smoother and more useful than those that came before. Lebesgue also worked on the theory of point sets, the calculus of variations, and dimension theory. He wrote a very large number of books and papers and had interests in the pedagogy and history of mathematics.

Lefschetz, Solomon. (b. Moscow, Russia, 3.9.1884; naturalized American citizen, 1913; d. Princeton, NJ, USA, 5.10.1972). Russian/American mathematician who greatly influenced three generation of American mathematicians. In an autobiographical note, written in 1968, he summarized his work: 'As I see it at last, it was my lot to plant the harpoon of algebraic topology into the body of the whale of algebraic geometry,' His approach was to pioneer the development of topological ideas at a time when algebraic geometry was being developed, notably by O. Zariski, along very algebraic lines. Many of the techniques he invented to deal with algebraic varieties have been applied successfully to complex manifolds; his technique of viewing a variety as made up of all its hyperplane slices is especially fertile. In algebraic topology he proved a very useful fixed-point theorem for transformations of manifolds which has inspired many profitable generalizations (*Algebraic Geometry and Topology*, Princeton, 1957). During WW2 he turned to the stability theory of non-linear differential equations and dynamical systems, which he redeveloped from a geometrical point of view, though with an eye to engineering applications. Lefscehetz was also a brilliant linguist who published articles and books in Russian, French, English and Spanish, and spoke many more languages including Persian. Originally at Kansas University, he spent his later years at Princeton.

Legendre, Adrien-Marie. (1752-1833), French mathematician who established many important results, especially in number theory and Elliptic Integrals, conjectured the Prime Number Theorem and the law of Quadratic Reciprocity, and published an elementary geometry textbook. He also published works on the orbits of comets and geodesy, and was appointed to a number of official positions.

Leibniz, Gottfried Wilhelm. (1646-1716), German philosopher, logician, mathematician and scientist. His initial academic training was in the law, and he left his home town of leipzig for ever when he was denied a doctorate of law there on the grounds that he was only 20; however, he was immediately awarded the degree at Nürnberg, where he was also offered a Chair. Refusing this offer, he instead entered the service of the Elector of Mainz, for whom he travelled widely; he was befriended by all the leading intellectuals of his age, and at this time developed the differential and integral Calculus independently of Newton. perfected Pascal's calculating machine, and laid the foundations of dynamics. He served the Duke of Hanover as everything from a schools inspector to a mining engineer (in which capacity he postulated the molten origin of the Earth), but he also continued his researches, describing the Binary System and laying the foundations of Topology. He postulated a perfect universal language of thought, in which logical relations would be transparent. He extended his duties as Historian to the House of Brunswick from genealogy to comparative linguistics and geology in pursuit of a universal history, and in this capacity he also influenced the succession of George I to Queen Anne of Britain. The son of a Lutheran professor of moral philosophy, Leibniz was educated at the universities of Leipzig, Jena, and Altdorf where he gained his doctorate in 1666. In 1667 he entered the service of the elector of Mainz for whom he spent the period 1673-76 on a diplomatic mission to Paris. Through meeting with such scholars as Christian Huygens in Paris and with members of the Royal Society, including Robert Boyle during two trips to London in 1673 and 1676, Leibniz was introduced to the outstanding problems challenging the mathematicians and physicists of Europe. On leaving Paris he joined the Staff of John Frederick the duke of Brunswick-Lüneburg, also Elector of Hannover, where he was given the commission to write the history of the House of Brunswick and the position of librarian. For the remaining 40 years of his life Leibniz dissipated his prodigious talents under three electors, including the future George I of Great Britain and Ireland, constructing genealogies of the numerous Brunswick progeny, both legitimate and, even more numerous, illegitimate. He also undertook a variety of administrative and diplomatic duties of which his attempt to unite the Protestant and Catholic churches in 1683 and the founding of the Berlin Academy of Sciences in 1700 are the most noteworthy. It was also to his Brunswick years that most of his philosophical writings belong although many of them remained unpublished until well after his death. Leibniz's greatest achievement was undoubtedly his discovery of the differential and integral calculus, work which was to involve him in a bitter priority dispute with Isaac Newton (q.v.). Newton's ideas on the calculus were developed first, as early as 1665, but remained unpublished until 1687; Leibniz, however, began work on problems of the calculus during his Paris years and published his results in 1684 in *Nova methodus pro maximis et minimis* (New Method for the Greatest and the Least). It was later suggested in 1699 that Leibniz's original inspiration may well have come from conversations in London in 1673 and in 1676 as well as letters of Newton to Henry Oldenburg shown to Leibniz. From this point the dispute became open to all and was conducted with considerable ferocity and not a little dishonesty. In fact, as became clear later on, the discoveries were made independently; the final triumph lay with Liebniz for it was his notation of differentiation and integration, rather than the fluxions of Newton, that have survived in modern textbooks. In physics, Leibniz's metaphysical principles also led him to deny Newtonian gravity acting at a distance on the grounds that: "A body is never moved naturally, except by another body which touches it and pushes it; after that it continues until it is prevented by another body which touches it. Any other kind of operation on bodies is either miraculous or imaginary." He also rejected Newtonian concepts of absolute space and time, arguing more plausibly that space was simply "the order of bodies among themselves while time was their order of succession. Leibniz also, with Huygens, developed the concept of kinetic energy. As well as his contributions to metaphysics and philosophy, Leibniz established the foundations of symbolic logic, probability theory, and combinatorial analysis, and was led to design a practical calculating machine. It was actually built and shown to the Royal Society in 1794.

L'Hopital, Marquis Guillaume Francois Antoine de. (b. 1661; Paris; d. Feb. 2. 1704; Paris). French mathematician L'Hôpital began his career as a cavalry officer but, forced to resign because of his short-sightedness, he devoted the rest of his life to mathematical study and research. To this end the invited the German mathematician Jean Bernoulli to his chateau in 1691 to teach him the details of his newly worked out differential calculus. Shortly afterward L'Hopital published his *Analyse des infiniment petits pour l'intelligence des lignes courbes* (1696), the first calculus textbook ever to appear. L'Hôpital's basic assumption was that " ...a quantity, which is increased or decreased only by in infinitely smaller quantity, may be considered as remaining the same." It was in this work that he first formulated the rule for finding the limiting value of a fraction with a numerator and denominator simultaneously tending to zero (0/0), since known to mathematicians as *L'Hôpital's rule.* Bernoulli appeared none too pleased with L'Hôpital's book considering it to be largely his own work, a belief supported by the discovery in 1921 of *Die Differentialrechnung,* a manuscript of Bernoulli's, on which L'Hôpital had clearly based his own text.

Lie, (Marius) Sophus. (b. Dec. 17, 1842; Nordfjordeid, Norway; d. Feb. 18, 1899; Kristiania, now Oslo) Norwegian mathematician. Lie was a friend of Felix Klein, whose ideas influenced him. Among Lie's most important work is his founding of the theory of continuous groups, which are now called *Lie groups* in his honor. Another contribution was his discovery of contact transformations. On both these subjects Lie wrote major treatises. In 1886 he became professor of mathematics at Leipzig and in 1898 he returned to Norway to take up a post that had been instituted for him at the University of Kristianina. By now, however, his health was poor and he died in Kristiania the following year. Lie also did notable work on differential geometry and on the study of differentiable equations.

Lin, Chia-Chiao. (b, July 7, 1916; Fujian, China) Chinese-American mathematician. Lin graduated from the National Tsing Hua University of Taiwan then obtained his MA from the University of Toronto in 1941 and his PhD in 1944 from the California Institute of Technology, Pasadena. He taught briefly at Brown University. Rhode Island (1945-47), before moving to the Massachusetts Institute of Technology. He was appointed professor of applied mathematics in 1953, becoming Institute Professor in 1966. Lin was worked on problems of hydrodynamics and turbulent flow in general. At a more particular level he has considered and is widely known for his account of how the spiral structure of spiral galaxies in sustained. Lin's account, known as the 'density-wave theory', is based on work by Bertil Lindblad and was worked out in collaboration with Frank Shu in 1964. They propose that the spiral structure is a rotating density wave that sweeps through the galaxy. The spiral pattern is always there but the material in the pattern is continuously changing under the influence of gravity. It is further supposed that as the spiral wave moves through a region it compresses the gas there sufficiently to trigger the process of star formation along the lines of compression. Young stars should therefore be found in the arms of spiral galaxy, as indeed they are. The model has received some additional observational support by predicting that spiral arms must trial and cannot lead as the galaxy rotates. Its universality has, however, been challenged by H. Cerola and P. Seiden who proposed in 1977 an alternative mechanism triggered by supernovae, without needing to assume the presence of an underlying density wave.

Lindemann, Carl Louis Ferdinand von. (b. Apr. 12, 1875; Hannover, now in West Germany; d. Mar, 1, 1939; Munich, now in West Germany). German mathematician. Lindemann is principally known for solving one particular mathematical problem, namely the question of whether or not the number π is transcendental. In his paper *Über die Zahl* π(1882; On the Number π) he showed that it is and thus disposed of the ancient problem of squaring the circle, i.e. wheather it is possible using only straight edge and compasses to construct a square equal in area to a given circle. Lindemann's result established its impossibility. Lindemann held posts as professor of mathematics at both the universities of königsberg and Munich. His other mathematical work was chiefly in analysis and geometry.

Liouville, Joseph. (b. Mar. 24, 1809; Saint-Omer, France; d. Sept. 8, 1882; Paris), French mathematician. Liouville was a professor at the Ecole polytechnique in Paris and edited an influential mathematical journal, the *Journal* *de mathematiques pures et appliquées*, often known as *Liouville's Journal*. The achievement for which he is best known in his work on transcendental numbers. Not only did he show that such numbers exit but he also devised methods of exhibiting infinitely many of them. Apart from this work on numbers, his other areas included differential equations and boundary-value problems. His work on boundary-value problems became part of what is now known as *Stürm-Liouville theory* and came to be of considerable importance in physics. Liouville also made contributions to differential geometry, in particular to the theory of conformal transformations, and to complex analysis of fundamental importance in statistical mechanics and measure theory.

Littlewood, John Edensor. (b. June 9, 1885; Rochester, England d. Sept. 6, 1977; Cambridge, England). British mathematician. Littlewood studied at Cambridge University, and in 1907 obtained a lectureship at Manchester. By 1910 he had returned to Cambridge where in 1928 he became Rouse Ball Professor of Mathematics. He retired in 1950 but continued active mathematical research untill shortly before his death. Littlewood is primarily known for his work in analysis, but he made contributions to many other fields, including mathematical astronomy, physics, differential equations, and probability theory. One of the most notable features of his career was his 35 year collaboration with G. H. Hardy. Among their most important joint work was the systematic investigation of problems of the convergence and summability of Fourier series. In 1914 Littlewood proved a famous theorem about the error term in the prime number theorem. In collaboration with R.E.A.C. Paley, Littlewood created and developed a new and specific link between trigonometric series and analytical functions. His work with Mary Cartwright on nonlinear differential equations was also of importance.

Lloyd Motz. Lloyd Motz was born in Susquehanna, Pennsylvania, in 1910. He received his baccalaureate at City College in New York in 1930 and his PhD at Columbia University in 1936. After working as a physics instructor at City College, he spent four years as director of research and optical design at Dome Precision Corporation. In 1946 he became a director of the Park Instrument Company. He worked as an associate professor at Columbia University from 1950-1962 before becoming professor of astronomy there. In 1960 he posited a new theory for the structure of fundamental particles that introduces gravitational fields into the interior of particles to account for their stability. He is currently professor emeritus at Columbia and engaged in a study of the structure of elementary particles using a gravitational model. His research has investigated the internal constitution of stars, unified field theory, the design of optical instruments, geometrical optics, and the structure of elementary particles. According to Motz's gravitational theory of elementary particles, a proton, like an electron, consists of three very massive particles held together gravitationally along a straight line. They then form a linear rotator or gravitationally bound triplet of basic particles spinning around a central axis. With his model of baryons he has explained the mass difference between the proton and neutron, the size of the proton and the magnetic moment (which is defined for a magnet as the strength of a magnet times its length) of the proton and neutron to within an accuracy of one percent. Motz has published many scientific papers and several astronomy textbooks including *Essentials of Astronomy* and *Astrophysics* and *Stellar Dynamics*. He has also written several books on astronomy and cosmology for the lay audience as well as edited various editions of Enrico Fermi's *Thermodyamics* and *Crystals, Molecules and Quantum Statistics.* This section draws upon the work of many of the authors in this section to present a comprehensive and nontechnical account of the "birth, evolution and death of stars." The Astronomer physicist and mathematician Lloyd Motz discusses the applicability of the second law of thermodynamics of the Universe as a whole. He qualifies its relevance with the point that we have no way of knowing whether or not the Universe is a closed system. It is possible that the Universe could interact with an outside system such as an alternate Universe. According to both the Einstein-Rosen and Kerr models, black holes could conceivably drain matter from this Universe and empty it into other Universes. Black hole in other Universes might also empty matter into this Universe and thereby maintain a general equilibrium. Yet we have no way of knowing whether the Penrose diagrams that mathematically describe the space-time of both static and rotating black holes are useful only in the abstract or have some physical relevance to the queer conditions existing inside the singularities of black holes. Motz also examines evidence that favors the pulsating or oscillating model such as indications that the rate of expansion is slowing down and could come to a complete halt in some 30 billion years, a time when the Universe will consist solely of burned-out stars, a few white dwarfs, neutron stars, and black holes. The gravitational collapse of the Universe could marry all black holes into a single cosmic black hole into which the entire Universe could disappear. However,

Motz believes that the quantum fluctuations of matter on the subatomic scale will probably prevent this cosmic black hole from collapsing to a point of zero volume and infinite density. The result of this blocked collapse would be a "bounce effect" which would begin a new cycle of expansion. Motz does not suggest the degree of uncertainty involved with predictions on the quantum level especially in a black hole since we cannot predict whether a new Universe would emerge in this space or if it would be inevitably crushed out of existence. Motz also devotes his attention to more localized scenarios for the end of the Universe as we know it including the sun's eventual evolution to a red giant after it has converted all its hydrogen to helium. The sun will maintain its bloated state for a relatively short time before contracting under its own gravitational field to become a white dwarf typical of the end stage of stars of its mass. In any case, the earth will be vaporized when the sun becomes a red giant. Even if the earth somehow escaped a fiery death, the solar white dwarf would give off so little light and heat that all life on earth would a quickly perish. Motz has also recently proposed a gravitational theory of the structure of elementary particles (protons, neutrons, and electrons) in which the gravitational force plays the dominant role and is thus the strong nuclear force. In this theory Motz identifies Gell-Mann's quarks (the constituents of protons and neutrons) with very massive particles that he calls unitons and whose existence he deduced theoretically from basic relativistic and quantum principles. Each uniton is 10^{19} times as massive as a proton, which means that when three unitons combine gravitationally to form a proton, they radiate away most of their mass. This vast release of energy accounts for the big bang if the Universe began as a cold, compact soup of unitons. The initial collapse of this soup would cause the unitons to combine gravitationally in triplets to form protons in an explosive release of energy which would initiate the expansion. The stray unitons left over in this big bang account for the hidden mass. According to this theory the Universe cannot collapse to a singularity because the collapse would be halted with the reappearance of unitons.

Lobachevsky, Nikolai Ivanovich. (b. Dec. 1, 1793; Nizhny Novgorod, now Gorky in the Soviet Union; d. Feb. 24, 1856; Kazan, now in the Soviet Union). Russian mathematician. Throughout his life Lobachevsky was associated with the University of Kazan. He was a student there and held various posts, including the chair in mathematics and finally the rectorship. Later he was deprived of his position for political reasons. Lobachevsky's fame is due to his epoch-making discovery, announced in 1826 and published in 1829, that there could be consistent systems of geometry based on other postulates than those of Euclid. In particular Labachevsky constructed and studied a type of geometry in which Euclid's parallel postulate is false (the postulate states that through a point not on a certain line only one line can be drawn not meeting the first line). Janos Bolyai had, at the same time though quite independently, come to a similar result and the same discovery had in fact been made decades earlier by Karl Friedrich Gauss, but he never published his work. For centuries the status of Euclid's geometry and in particular of his parallel postulate had been a matter of controversy. Attempts had been made to show that it followed from the other axioms and it was widely held that Euclid's geometry described the necessary structure of space. By revealing the coherence of a non-Euclidean geometry Lobachevsky showed that Euclidean geometry has no such privileged position, helped mathematicians to break free from undue reliance on intuition, and paved the way for the systematic study of different kinds of non-Euclidean geometry in the work of Bernhard Riemann and Felix Klein. Although it was not well received at first the value of Labachevsky's work was fully appreciated once Riemann began his investigations into the fundamental concepts of geometry. Perhaps his fullest vindication came with the advent of Einstein's theory of relativity when it was demonstrated experimentally that the geometry of space is not described by Euclid's geometry. Apart from geometry, Lobachevsky also did important work in the theory of infinite series, algebraic equations, integral calculus, and probability.

Lotka, Alfred James. (b. Lemberg, Galicia, Austria-Hungary [now Lvov, Ukraine, USSR], 2.3.1880; d. Red Bank, NJ, USA, 5.12.1949). American mathematician. Lotka was born of American parents; he was educated in the UK at Birmingham University, and then at Cornell. For most of his life he worked in New York for the Metropolitan Life Insurance Company. His main work was in the application of mathematical theory.to biological situations; he was particularly concerned with evolution and the dynamics of human and animal populations, modelled in terms of birth and death rates, and age distribution. He believed that 'the logistic (curve) expresses what may be regarded as a fundamental law of populations, that a population cannot increase indefinitely in a constantly geometric progression.' His major book *Elements of Physical Biology* (NY, 1925) was reprinted in 1956 as *Elements of Mathematical Biology*. In 1932

Lotka published a paper entitled 'The growth of mixed populations: two species competing for a common food supply'. This was a development of mathematical principles published earlier by Volterra. By presenting a growth law for two competing populations, expressed in terms of interlocking differential equations, Lotka gave a theoretical exposition of the principle of competitive exclusion in the same year that Gause offered a description of the principle based on experimental work.

Love, Augustus Edward Hough. b. Apr. 17, 1863; Weston-super-Mare, England, d. June 5, 1940; Oxford, England). British mathematician and geophysicist. Love studied at Cambridge University and was a fellow there from 1886 until 1889. In 1889 he moved to Oxford to take up the Sedleian Chair in Mathematics and remained there for the rest of his career. Love's most important research was in the theory of deformable media and in theoretical geophysics. He also did important work in the theory of waves and in ballistics. He wrote a major two-volume treatise on elasticity—*A Treatise on the Mathematical Theory of Elasticity* (1892-93)—that went through several subsequent editions. It soon established itself as a classic and is still a standard work of reference. Love's work on geophysics led to a considerable number of practical discoveries. Among his original concepts now much used in geophysics are the so-called *Love numbers* and *Love waves.* Love numbers play a key role in tidal theory. In the formal theory of Love waves, which are surface seismic waves, he was building on and much improving work begun by Siméon-Denis Poisson, Sir George Stokes, and Lord Rayleigh. This was perhaps Love's single most significant piece of mathematical work.

Lubbock, John William. (b. Mar. 26, 1803; London; d. June 20, 1865; Farnborough, England). British mathematical physicist. Lubbock was the son of a banker who also had scientific interests. He was educated at Eton and Cambridge University, England, where he graduated in 1825 and then became a partner in his father's bank. During the period 1813-36 Lubbock published a number of tidal studies. By working on the tidal records of the London docks from 1795 he was able to work out the establishment of port, the time high water falls behind the Moon at a particular place on the Earth. He was much influenced by Pierre Simon de Laplace whose innovations in probability theory he used in his theoretical studies of annuities and whose work on the stability of the solar system he advanced. Lubbock was the first vice-chancellor of the University of London (1837-42).

Ludwig Boltzmann. Ludwig Boltzmann was an Austrian physicist whose career, of rare brilliance, was cut short by his suicide in 1906 at the age of sixty-two, an act committed during a period of severe mental depression that was possibly aggravated by opposition to certain of his ideas on atomic structure by some of the outstanding chemists of the time. He developed a version of the kinetic theory of gases quite independently of James Clerk Maxwell. These two scientists are usually given equal credit for this great advance in physics. Boltzmann also helped to create the discipline of statistical mechanics by subjecting Clausius' theory of increasing entropy (that is, the second law of thermodynamics) to strict mathematical treatment. As a young student Boltzmann had been an assistant to the physicist Josef Stefan, whose research on the cooling of hot materials culminated in his law stating that the total radiation given off by a hot body is proportional to the fourth power of its absolute temperature (that is, if the absolute temperature is doubled, the radiation rate increases sixteen times). In 1984 Boltzmann proved that the same principle could be derived from thermodynamics, and thus this important physical tenet is usually referred to as the Stefan-Boltzmann law. Boltzann was convinced that all physics could be explained in terms of atomic processes. He applied Maxwell's statistical techniques to large numbers of atoms, eliminating the need to make the often questionable assumption that all atoms move at about the same velocity and are equally spaced. Maxwell had previously described the billions of atoms in a gas not by billions of coordinates but by one quantity known as the distribution function. This distribution function was represented by a table divided into boxes which included in sum all the different combinations of positions and velocities a molecule can have. These combinations were divided into small ranges, and each box in a particular range gave the average number of molecules within that range. The distribution function does not give the coordinates of any one atom, but for a particular range it will give the probability of finding an atom in that range. From the distribution function the summed or average values of all atomic quantities can be found. Boltzmann discovered the law that governs how this distribution evolves over time under the influence of collisions among the myriads of atoms and applied it successfully to the solution of many problems. Yet his work in atomic distribution was attacked by "energeticists" like Ernst Mach, who accepted atoms only as convenient mathematical fictions but not as physically extant objects, and Georg Helm, who denied both their existence and their convenience. The energeticists presumed everything to be made of energy and took the law of conservation of energy to be the supreme law of nature. The strongest arguments against atoms lay in the paradoxes that seemed to follow from the atomic theory of matter. Maxwell had already noted the paradox of reversibility. Thermodynamics has a preferred direction for physical processes, such as ice melting on a hot stove, yet all mechanical processes, such as the orbiting of a planet around the sun, can go equally well in either direction. Maxwell had also stated the return paradox: Any mechanical system of atoms in a finite volume like a box will return to any starting point after a long but finite time. If the ice cube and the water in a cold drink are a mechanical system of atoms, the melting ice will refreeze if we wait long enough for the atoms to return to their original positions. Boltzmann spent a great deal of time thinking about these paradoxes because they dealt directly with the existence of atoms. He concluded that these so-called paradoxes are actually true statements consistent with observation. The irreversibility of thermodynamics is entirely a matter of initial conditions. If we could prepare these conditions correctly, an isoinitial conditions. If we could prepare these conditions correctly, an isolated system would indeed reverse its thermodynamic evolution. The return time, however, is many times longer than the lifetime of the Universe. Boltzmann's supreme accomplishment was to link the central quantity of thermodynamics, entropy, to the statistics of atoms. His formula, one of the most enduring in physics, is also among the simplest: $S = k\ln W$. The equation states that the entropy *(S)* of a system measures the probability (W) of its state The variable *k* is a fundamental new physical constant now called Boltzmann's constant and "In" stands for a natural logarithm. The left - hand side of the equation is a creature of thermodynamics, captured solely with such macroscopic snares as calorimeters and thermometers. The right - hand side is a denizen of the microscopic world, a product of atomic statistics. This equation enabled Boltzmann to translate the language of atoms into the common vernacular of thermodynamics.

Lukasiewicz Jan. (b. Lemberg, Galicia, Austria-Hungary [now Lvov, Ukraine, USSR], 21.12.1878; d. Dublin, Irish Republic, 13.11.1956). Polish logician. Lukasiewicz was one of the founders of the remarkable school of logicians which sprang up in Poland after WWI. He made important contributions to modal logic (in which 'is possible' is allowed as a fundamental notion) and many-valued logics (in which truth-values additional to 'true' and 'false' are admitted). He hoped that such logics would become accepted into logic in the same way that non-Euclidean geometries were taken into mathematics in the mid-19c. He also introduced into logic the 'Polish notation' which has found favour in computing for its avoidance of brackets. Lukasiewicz left his chair at the University of Warsaw near the end of WW2. In 1946 he was appointed to the Royal Irish Academy in Dublin, where his continuing studies included and analysis of Aristotle's logic in terms of the recent discoveries in axiomatization and deduction. His book *Aristotle's Syllogistic from the Standpoint of Modern Formal Logic* (Oxford, 1951,[2] 1957) has inspired a large collection of such studies. Some of his papers are published in *Selected Work* (Amsterdam, 1970).

M

Mach, Ernst (b. Feb. 18, 1838; Turas, now in Czechoslovakia; d. Feb. 19, 1916; Haar, near Munich, now in West Germany). Austrian physicist. Mach had a somewhat unorthodox upbringing and education. His father was a man knowledgeable in both the classics and the sciences who retired to farm near Vienna where he educated his son academically, but also emphasized the importance of such practical skills as carpentry and farming. He received his higher education from the University of Vienna obtaining his PhD there in 1860. He was appointed professor of mathematics at Graz in 1864 and in 1867 was appointed to the chair of physics at the University of Prague, moving to Vienna in 1895 where he became professor of the history and theory of the inductive sciences. He was forced to retire in 1901 after being partially paralyzed by a stroke. Mach made a large number of contributions to science in a variety of field, but it is as a critic of science and as a philosopher that he exercised such a powerful influence over several generations of scientists. He began his research career in experimental psychology, inspired by the work of such scholars as Gustav Fechner. For several years he investigated various perceptual processes; in the course of this work he discovered the function of the semicircular canals of the ear. He is however best known to a wider public for his work on shock waves, which eventually led to the *mach number* being introduced (1929) as a measure of speed (it is the ratio of the speed of an object in a fluid to the speed of sound in the fluid). Mach's interest in shock waves was aroused by a claim made by L. Melsen that the savage wounds of those shot in the Franco-Prussian war were not due to the ammunition used by the French but by the impact of compressed air pushed ahead of the bullet. In 1884 Mach began his attempt to photograph the shock and sound waves produced by projectiles. It was not until 1886 that he was successful and able to conclude that Melsen was wrong. Mach is also known to cosmologists for his controversial statement of the principle of inertia, called *Mach's principle* by Einstein over whom Mach exerted considerable influence. For Newton the inertia of a body was an intrinsic property independent of the presence or absence of any other matter. For Mach the concept of inertia could only acquire quantitative meaning through Newton's laws. But Newton formulated his laws against the background of absolute space and time. This latter point Mach rejected completely, claiming that only relative motion exists. Thus, for Mach, it should make no difference whether we talk of the Earth rotating on its axis relative to the fixed stars or whether we see the Earth as fixed and talk of the stars as moving relative to the Earth. There is however, the significant difference that only in the first case can the concept of inertia be used to understand such phenomena as the flattening of the Earth and Foucault's pendulum. But Mach was a radical thinker and argued that the inertia is not an intrinsic property of matter but is itself caused by the background of the fixed stars. If these were removed inertia would disappear with them. Mach's elimination of absolute space was simply part of a more general program in which he hoped to eliminate metaphysics — all those purely 'thought — things' which cannot be pointed to in experience — from science. He began by asserting that the world contains nothing but sensations and their connections; scientific laws describe such connections in the simplest possible way; they provide an 'economy of thought. His views influenced the important philosophical movement of logical positivism and also had some impact on scientific practice, influencing Einstein in his theory of relativity. There is no doubt however that his approach could lead him astray. He never, for example, accepted the existence of atoms. They were rather, "economical ways of symbolizing experience. But we have as little right to expect from them, as from the symbols of algebra, more than we have put into them. "Nor did he ever accept the relativity theory of Einstein.

Maclaurin, Colin. (1698-1747), Scottish mathematician and physicist, who developed Newton's work in both subjects. He entered Glasgow University at age 11, was appointed Professor of Mathematics at Aberdeen at 19, was elected to a Fellowship of the Royal Society at 21, moved to the Chair at Edinburgh on Newton's recommendation at 27, and in 1740 shared a prize offered by the French Academy of Science for an essay on tides with Euler and Daniel Bernoulli; he also wrote a defence of Newton's theory of FLUXIONS against the philosopher Berkeley. During the 1745 rebellion he was active in preparing the defences of Edinburgh against Bonnie Prince Charlie, and had to flee to England when the city cell to the rebels; although he returned after their swift defeat, his health never recovered and he died the following year. Maceaurin was a child prodigy; he entered Glasgow University at the age of 11 and became professor of mathematics at marischal College, Aberdeen, at the age of 19. His chief work was Geometriea *Organica, sive descriptio linearum curvarum universalis* (1720; Organic Geometry, with the Description of the Universal Linear Curves) and in this he proposed several theorems that developed along similar lines to those contained in Isaac Newton's *Principia.* Maclaurin became a friend of Newton and defended Newton's new theory of calculus against the polemics of the Irish philosopher George Berkeley. Maclaurin was one of the first to treat the theory of maxima and minima properly. He also contributed to the theory of the equilibrium of rotating bodies of fluid. The *Maclaurin expansion*, which is a special case of the Taylor series, was named for him. Maclaurin played an important role in organizing the defense of Edinburgh against the Jacobites in 1745 and when they captured the city he was forced to flee to England.

MacMahon, Percy Alexander. (b. Sept. 26, 1854; Malta; d. Dec. 25, 1919; Bognor Regis, England) British mathematician. MacMahon was a mathematical coach at the Royal Military Academy from 1822 to 1888. Subsequently he was assistant inspector at Woolwich Arsenal (1888-91), and professor of physics at the Artillery College (1891-98). His chief mathematical work was in classical algebra, in particular in combinatorial analysis. He built on the work of Arthur Cayley and James Joseph Sylvester, and thus belongs squarely in the 19th century British school of algebraists. MacMahon was particularly interested in symmetric functions. He distilled his ideas on combinatorial analysis into a book that became a classic on the subject.

Markov, Andrey Andreyevich. (b. June 14, 1856; Ryazan, now in the Soviet Union; d. July 20, 1922; Petrograd, now Leningrad in the Soviet Union). Russian mathematician. Markov studied at the University of St. Petersburg and later held a variety of teaching posts at the same university, eventually becoming a professor in 1893. He was an extremely keen and effective teacher. His mathematical interests were very wide, ranging over number theory, the theory of continued fractions, and differential equations. It was, however, his work in probability theory that constituted his most profound and enduring contribution to mathematics. Among Markov's teachers was the eminent Russian mathematician Pafnuti Chebyshev, whose central interest was in probability. One of Markov's first pieces of important research centered on a key theorem of Chebyshev's — 'the central limit theorem'. He was able to show that Chebyshev's supposed proof of this result was erroneous, and to provide his own, correct, proof of a version of the theorem of much greater generality than that attempted by Chebyshev. In 1900 Markov published his important and influential textbook *Probability Calculus,* and by 1906 he had arrived at the fundamentally new concept of a *Markov chain.* A sequence of random variables is a Markov chain if the two probabilities conditioned on different amounts of information about the early part of the sequence are the same. This aspect of Markov's work gave a major impetus to the subject of stochastic processes. The great importance of Markov's work was that it enabled probability theory to be applied to a very much wider range of physical phenomena than had previously been possible. As a result of his work a whole range of subjects, among them genetics and such statiscal phenomena as the behavior of molecules, became amenable to mathematical probabilities treatment.

Max Born. Max Born Shared the 1954 Nobel Prize in physics with Walter Bothe for his statistical studies of wave functions. These studies represented an important emendation of quantum theory in which electrons are regarded as particles whose behavior can be mathematically described with great precision. Born was the teacher of Werner Heisenberg and worked with many of the leading physicists on the Continent and in Great Britain during the 1920s and 1930s. Born was born in Breslau in 1882. His father was an anatomist and embryologist and his mother was a member of a Silesian family of industrialists. He studied at the Universities of Breslau, Heidelberg, Zurich, and Göttingen. At Göttingen he studied mathematics with Klein, Hilbert, Minkowski, and Runge and astronomy with Schwarzschild before graduating in 1907. After spending several years at Cambridge and Breslau, Born became an academic lecturer at Göttingen until moving on to the University of Chicago for several years to do spectroscopy experiments with Michelson. Despite a 1915 appointment as professor in Berlin, Born was drafted into the German army and spent the duration of the war in a science office working on his theories of sound and crystals. After the war he returned to Göttingen, where he did his most important work in quantum theory. Born left Germany in 1933 because of Nazi racial laws (he was Jewish) and sought asylum in

England. He was appointed Stokes Lecturer at Cambridge University and became a British citizen in 1939. After retiring in 1953 he returned to his birthplace, Göttingen, where he lived until his death in 1970, writing articles and books, many of which dealt with the responsibilities of scientists in the nuclear age. The selection by Born, "On the Meaning of Physical Theories," is the text of a lecture given before a Göttingen academy in 1928. In it Born makes a trenchant analysis of the value of theory in physics and the confusion that results because at times "what today is valuable knowledge will tomorrow be so much junk, hardly worth a historical backward glance. "In making his point he roams over vast stretches of the history of physics, raging from Lucretius to Heisenberg, scarcely missing an important branch of physics in what is virtually an inspired an highly readable survey course in classical and modern physics, including an unexpected reference to William Tell (not in connection, however, with the uncertainty principle.

Max Planck. Altough Max Planck, took his doctorate *summa cum laude* in physics under Helmholtz, Clausius, ad Gustav Kirchhoff (the celebrated pioneer of spectroscopy), he seems not to have impressed his illustrious masters to any remarkable degree. He wrote his thesis on thermodynamics, but he was later heard to complain that Helmholtz never bothered to read it and that, although Kirchhoff read it, he apparently did not think much of it. Even so, Kirchhoff's interest in black body radiation led Planck to study that phenomenon and to undertake the research that led to his theory of quanta and to the founding of an important part of twentieth century physics. In his study of black body radiation, Planck found that two of his predecessors, Whilhelm

Wien and John William Strutt, Lord Rayleigh, had derived equations to describe the distribution of radiation over the entire range of frequencies emitted by a black body. He also found that Wien's equation works fine for high frequencies but not for low frequencies, while Rayleigh's equation is applicable to low frequencies but not to high frequencies. Almost in sheer desperation, Planck later declared, he devised a simple equation by algebraically combining Wien's and Rayleigh's equations into one that worked for the entire spectrum, basing it on the assumption that all energy, like matter, exists in small entities, to which he gave the name *quanta* (singular, *quantum,* from the Latin term for "how much?". He announced his discovery, or rather his hypothesis, at a famous meeting of the Berliner Physikalische Gesellschaft on 14 December 1900, reading a paper on his work for which he was awarded the Nobel Prize eighteen years later. Planck's interest in entropy began with extensive studies of Clausius' papers, which inspired in him an unshakable belief in the eternal truth of the second law of thermodynamics and caused him at first to favor the "second" (kinetic) theory of heat over that "first" (caloric) theory. Planck set out to do for entropy what Helmholtz had done for energy: extend it to encompass all the forces of nature, beginning with the antropy of elastics solids and going on to the entropy of electric and magnetic fields and light. Although he defended Calusius against the "energetics" of Ernst Mach and Georg Helm, he ultimately turned to a "third" theory of heat which obtained completely from any assumption about the microscopic nature of heat but relied entirely on macroscopic concepts.

Maupertuis, Pierre-Louis Moreau de. (b. Sept. 28, 1698;

Saint-Malo, France; d. July 27, 1759; Basel, Switzerland). French mathematician, physicist, and astronomer. Maupertuis joined the army as a youth, leaving in 1723 to teach mathematics at the French Academy of Sciences in Paris. He traveled to England in 1728 where he became an admirer of Isaac Newton's work and was made a member of the Royal Society of London. He was responsible for introducing Newton's theories on gravitation into France on his return. In 1736 Maupertuis led an expedition to Lapland to verify Newton's hypothesis that the Earth is not perfectly spherical by measuring the length of a degree of longitude. This was successful and as a result Maupertuis was invited by Frederick the Great to join his Academy of Sciences in Berlin. Maupertuis is best known as being, in 1744, one of the first to formulate the principle of least action, which was published in his *Essai de cosmologie* (1750; Essay on Cosmology). A similar principle had previously been formulated by Leonhard Euler as a result of his mathematical work on the calculus of variations, whereas Maupertuis had been led to formulate his version of the principle through his work in optics. In particular Maupertuis's attention was drawn to the need for such a principle by his interest in the work of Willebrord Snell's and Pierre de Fermat. Fermat had shown how to explain Snell' law of refraction, which describes the behavior of a ray of light at the boundary of two media of different densities on the assumption that a ray of light takes the least time possible in traveling from the first medium to the second. However, Fermat's explanation implied that light travels more slowly in a denser medium and Maupertuis set out to devise an explanation of Snell's law that did not have this, to him, objectionable consequence. Maupertuis thought of his principle as the fundamental principle of mechanics, and expected that all other mechanical laws ought to be derivable from it. He attempted to derive from his principle a proof of the existence of God. Maupertuis was not a mathematician of Euler's stature and his version of the principle was not as precisely formulated mathematically. However, his attempts to apply it to a much wider range of problems made it an influential formative principle in 18th-and 19th-century physical thinking. Jpseph Lagrange, in his work on the calculus of variations, dispensed with the teleological and theological trimmings Maupertuis had given the principle and found wide application for it in mechanics. Subsequently the principle became less influential until it was revived and refined by William Hamilton. Maupertuis, who had a quarrelsome character, became involved in violent controversy over the principle. Samuel König, another scientist at Frederick's court, claimed that it had been formulated earlier by Gottfried Leibniz. Maupertuis found himself on the receiving end of some of Voltaire's most biting satires, and eventually he was hounded out of Berlin.

Mengoli, Pletro (b. 1625; Bologna, Italy; d. 1686; Bologna). Italian mathematician. Mengoli studied in Bologna with Francesco Cavalieri. He was ordained a priest and held a chair in mathematics at Bologna. The methematical work was chiefly on the calculus and infinite series and his most important work was with convergence and with integration. He was on of the first to establish results about the conditions under which infinite series converge. His work constitutes an important stage in the development of the fully fledged calculus of Isaac Newton and Gottfried Leibniz from such ideas as

Cavalieri's method of indivisible. Mengoli gave a definition of definite integral, which to some extent anticipates Augustin Cauchy's.

Mersenne, Marin (b. Sept. 8, 1588; Oize, France; d. Sept. 1, 1648; Paris). Franch mathematician, philosopher, and theologian. Mersenne was educated at the famous Jesuit college at La Flèche. After this he studied theology at the Sorbonne in Paris and became a Catholic priest in the Order of Minims in 1611. He spent most of the rest of his life teaching at their convent near Place Royale but also traveled extensively in Europe. Despite his commitment to the Catholic faith, Mersenne took a very lively interest in the rapidly developing world of the physical sciences seeing his task as one of using the new discoveries as a means to defend Catholic orthodoxy. Probably his most important contribution to science was his role as a communication link between scientists. He kept in contact with as many eminent scientists of the day as possible, as his enormous correspondence testifies, and there were few 17th-century men of science of any importance who did not correspond with Father Mersenne. His convent at Place Royale became a regular meeting place for what were in effect conferences of leading scientists and philosophers. Among those who gathered here were Blaise Pascal, René Descartes, Pierre de Fermat, and the philosopher Thomas Hobbes. Mersenne himself was particularly interested in problems of scientific methodology. He doubted the range and usefulness of the Euclidean axiomatic method advocated for the sciences, especially by Descartes. Mersenne was a vigorous opponent of the various forms of skepticism that were popular at the time, and was much interested in the viability of the new mechanistic world picture which the new science seemed to bring with it. Mersenne's interests extended to nonscientific aspects of philosophy and he was one of the first to try to popularize the idea of an invented universal language. His interest in science was by no means purely theoretical and he did important practical work in optics and acoustics. Mersenne proposed, in 1644, his *Mersenne numbers*. These are numbers generated from the formula 2^p-1, in which p is a prime. This formula does not represent all primes but it contributed to developments in the theory of numbers.

Milne, Edward Arthur. (b. Feb. 14, 1896; Hull, England; d. Sept. 21, 1950; Dublin). British mathematician and astrophysicist. Milne, the son of a headmaster, studied at Cambridge University, 1914-16. He returned there in 1919 after working on ballistics during World War I, and was appointed as assistant director of the Cambridge Solar Physics Observatory in 1920. He lectured in both mathematics and astrophysics. In 1924 he became professor of applied mathematics at the University of Manchester where he remained until 1928. In 1929 Milne was appointed professor of mathematics at Oxford, a post he held, apart from a period working for the Ordnance Board (1939-44), until his death in 1950. Milne's wartime work had involved studies of the Earth's atmosphere. From 1920 he extended this to theoretical research on the atmospheres of stars, concentrating on the flow of radiation through these outermost layers of stars and on the ionization of the component atoms. In collaboration with Ralph Fowler, Milne used fix the temperature scale for the known sequence of stellar spectra. Thus just by knowing that a star falls into spectral type G permitted its temperature (5000-6000 kelvin) to be inferred. Following three years' research into stellar structure Milne turned in 1932 to the development of his theory of 'kinematic relativity'. This was an alternative to Einstein's theory of general relativity and contained a new cosmological model of the universe from which he evolved new systems of dynamics and electrodynamics. It was Milne who introduced the 'cosmological principle' that simply states that the universe appears essentially the same from wherever it is observed. This is still the basis of much of modern cosmology and became for Milne one of the axioms of an axiomatic cosmology that he hoped to construct in a purely deductive way. "Starting from first principles (like Descartes)" his aim was to "pursue a single path towards the understanding of this unique entity the universe; and it will be a test of the correctness of our path that we should find at no point any bifurcation of possibility." Our path should nowhere provide any alternatives" However, Milne did find alternative universes littering his path and in each case he behaved as Descartes had done before him and selected that alternative he favored as being the only one compatible with the rationality of the Creator. Milne's world model, which is now seen as a misguided curiosity, led nowhere. But for his early death he might well have developed it and brought it more into line with orthodox cosmology.

Minkowski, Hermann (b. June 22, 1864; Alexotas, now in the Soviet Union; Jan. 12, 1909, Göttingen, now in West Germany). Russian-German mathematician. Minkowski's parents were of German origin and the family returned to Germany in 1872, settling in Königsberg (now Kaliningrad). Minkowski studied alongside David Hibert at the University of Königsberg, under Adolf Hurwitz, and gained his PhD in 1885. He taught at Bonn (1885-94) and Königsberg (1894-96) and then worked with Hurwitz at the Zurich Federal Institution of Technology (1896-1902). At Hilbert's instigation a new chair of mathematics was created for Minkowski at the University of Göttingen and he worked there (1902-09) until his death. In 1883, when still 18, Minkowski was awarded the Grand Prix des Sciences Mathematics of the Paris Academy of Sciences. The award was shared with Henry J. Smith for their work on the theory of quadratic forms. Minkowski remained occupied with the arithmetic of quadratic forms for the rest of his life. In 1896 he gave a detailed account of his 'geometry of numbers' in which he developed geometrical methods for the treatment of certain problems in number theory. During his short period at Göttingen Minkowski worked closely with David Hilbert and decisively influenced Hilbert's interest in mathematical physics. Minkowski's most celebrated work was in developing the mathematics that played a crucial role in Einstein's formulation of the theory of relativity. Einstein knew when he published the special theory of relativity in 1905 that the universe could not be adequately described using normal, or Euclidean, three-dimensional geometry. Minkowski's seminal idea was to view space and time as forming together a single four- dimensional continuum or manifold, known as 'space-time', rather than two distinct entities. In normal three-dimensional geometry, any point in space can be identified by three coordinates. The analog of this point in three-dimensional space in an event localized both in space and time in four-dimensional space-time. Minkowski put forward his concept of space-time, or *Minkowski space* as it is sometimes called, in 1907 in his book *Space and Time*. Einstein himself was very forthright about the extent to which the theory of relativity depended on Minkowski's innovatory work. Space-time was a useful and elegant format for special relativity, and was essential for general relativity, published in 1916, in which space-time is allowed to be curved. It is the curvature of space-time that accounts for the phenomenon of gravitation.

Möbius, August Ferdinand. (b. Nov. 17, 1790; Schulpforta, now in East Germany; d. Sept. 26, 1868; Leipzig, now in East Germany). German mathematician. Möbius's main work was in analytical geometry, topology, and theoretical astronomy. He held a chair in theoretical astronomy at Leipzig and made numerous contributions to the field with publications on planetary occultations ('eclipses') and celestial mechanics. His more purely mathematical work centers on geometry and topology. Mobius is chiefly famed for his discovery of the *Möbius strip,* a one-sided surface formed by giving a rectangular strip a half-twist and then joining the ends together. He introduced the use of homogeneous coordinates into analytical geometry and did significant work in projective geometry, inventing the *Möbius net,* which became of central importance in the future development of the subject.

Monge, Gaspard. (b. May 10, 1746; Beaune, France; d. July 28, 1818; Paris). French mathematician. Monge was trained as a draftsman at Mézières where he later became professor of mathematics (1768). During the French Revolution he served on the committee that formulated the metric system (1791), became minister of the navy and the colonies (1792-93), and played a vital part in organizing the defense of France against the counterrevolutionary armies. He contributed significantly to the founding of the Ecole Polytechnique in 1795. Monge met Napoleon in 1796 and saw active service in Napoleon's army during the Egyptian campaign (1798-

1801). Monge's major mathematical achievements were the invention of descriptive geometry and in his application of the techniques of analysis to the theory of curvature. The former is indispensable in mechanical drawing and the latter ultimately led to the revolutionary work of Georg Riemann on geometry and curvature. Monge never changed in his antiroyalist sentiments and as a result, following Napoleon's fall from power in 1815, he was expelled from the French Academy and deprived of all his honors.

Mordell, Louis Joel. (b. Jan. 28, 1888; Philadelphia, Pennsylvania; d. Mar. 6, 1972; Cambridge, England). American-British mathematician. Mordell had his first mathematical education at school in Philadelphia. He decided to try for a scholarship to Cambridge University, England, and having scraped together enough money for a one-way ticket made the crossing, took the exam, and got the scholarship. Mordell taught at Birkbeck College, London, from 1913 to 1920. From 1920 to 1945 he held posts at the Manchester College of Technology and at Manchester University. Whilst in Manchester he got to know Sydney Chapman, and the two remained lifelong friends. From 1945 to 1953 he held the Sadleirian Chair in Mathematics at Cambridge. Mordell's central interest was in the theory of numbers and, in particular, Diophantine equations. He worked on the theory of modular functions and their applications to the theory of numbers. In the 1020s he published his most important single result, the Mordell finite basis theorem. André Weil later generalized it, and the result, which plays a fundamental role in number theory, is now usually known as the *Mordell-Weil theorem*. Among his other works are results on the estimation of trigonometric and character sums, cubic surfaces and hypersurfaces, and the geometry of numbers.

N

Napier, John. (b. 1550; Edinburgh; d. Apr. 4, 1617; Edinburgh). Scottish mathematician. Napier studied at the University of St. Andrews but left before taking his degree and then traveled extensively throughout Europe. He was a fervent Protestant and wrote a diatribe attacking Catholics and others whose religious views he disapproved of Napier was also very active in politics and he designed a number of war-engines of various kinds when it was believed that the Spanish were about to invade Scotland. Napier devoted his spare time to mathematics, in particular to methods of computation. He introduced the concept of logarithms, publishing his work on this in *Mirifici logarithmorum canonis descriptio* (1614; Description of the Marvelous Canon of Logarithms). Napier's tables used natural logarithms, i.e. to base e, and soon after their publication the tables were slightly modified by Henry Briggs to base 10. Napier's further work on logarithms was published after his death in *Mirifici logarithmorum canonis constructio* (1619; Construction of the Marvelous Canon of Logarithms). Napier did some other mathematical work, in particular in spherical trigonometry and in perfecting the decimal notation.

Newton Isaac. With the possible exception of Albert Einstein, no scientist so altered the human perception of the

Universe as Isaac Newton. His cosmos of absolute space and time drew upon the work of others such as Kepler and Galileo but it was Newton who brought together Galileo's mechanics and Kepler's laws of planetary motion to fashion a Universe which could run without the benefit of continual divine intervention. The elegance and order of the Newtonian system impressed even the most vocal of his critics. Although only a handful of people actually understood Newton's writings, his achievements were hailed not only by other scientists but also by some of England's most distinguished literary figures. Indeed, some of these remarks have been so overused as to become timeworn clichés, but they do help to illustrate the tenor of the praise given to Newton by both scientific and layman audiences. The first cliché is provided by Alexander Pope, who captured the adulation that was accorded Isaac Newton in his own lifetime in recognition of his theories and discoveries (which arguably wrought as great a change in our perception of the Universe as Copernicus's had a century and a half before) when he wrote:Nature and Nature's laws lay hid in night:God said, *Let Newton be!* and all was light. The second cliché is given by Isaac Newton himself. Writing to Robert Hooke in 1676 (before they became bitter enemies), he remarked:If I have seen further than other men, it is because I stood on the shoulders of giants.... Newton recognized that earlier scientists such as Kepler and Galileo had played crucial roles in laying the groundwork for his system of mechanics. Kepler's desire to describe the Universe mathematically had inspired him to write in 1605 "that the celestial machine is to be likened not to a divine organism but rather to a clockwork." Newton preferred to view the Universe as a gigantic watchspring wound up by the hand of God and left to run down on its own. Yet Kepler's success in explaining the orbits of the planets helped to convince Newton that the celestial mechanics of the Universe was governed by a number of unchanging mathematical laws. It remained for Newton to discover these laws or at least devise a framework to explain how the world works. The linchpin of Newton's efforts was to be a precise language of mathematical relationships which would make possible the predictable and systematic analysis of the motions of all celestial bodies. Daniel J. Boorstin writes that this breakthrough, made possible in part by Newton's ability to reconcile and synthesize the accomplishments of his predecessors, made it possible to unify and explain a wide range of seemingly unrelated phenomena: The power and grandeur of Newton's system consisted, of course, in its universality. He finally offered one common scheme for terrestrial and celestial dynamics. He had brought the heavenly bodies down to earth, and at the same time provided a new framework, and new limits, for man's grasp on the heavenly bodies.... He unified all the physical phenomena on the earth with those in the heavens by the generality of his laws, expressed mathematically. For all the motions of earthly and heavenly bodies could be seen, observed, and measured. The grand unifying force in Newton's system, even before gravitation, was mathematics.* It is truly difficult to grasp the impact that Newton's works, especially his *Principia*, had on his century. Yet he was often criticized by natural philosophers because his analytical efforts were geared to describing fundamental rules of order, not to explaining what causes all physical processes to occur. Newton never presumed that his world system was the apex of human ingenuity from which no further advances could occur. Nor was he so bold as to suggest that his work could somehow explain the motivations of the Creator. Indeed, Newton believed that it was impossible for science to transcend the barrier between our world and the spiritual world and was quite content to leave such speculations to other natural philosophers. In describing the limits of his method's usefulness, Newton wrote:... [W]e have explained the phenomena of the heavens and of our sea by the power of gravity, but have not yet assigned the cause of this power.... I have not been able to discover the cause of those properties of gravity from phenomena, and I frame no hypotheses; for whatever is not deduced from the phenomena is to be called an hypothesis; and hypotheses ... have no place in experimental philosophy. In this philosophy particular propositions are inferred from the phenomena, and afterwards rendered general by induction. Thus it was that the impenetrability, the mobility, and the impulsive force of bodies, and the laws of motion and of gravitation, were discovered. And to us it is enough that gravity does really exist, and acts according to the laws which we have explained, and abundantly serves to account for all the motions of the celestial bodies, and of our sea. It is a pleasing coincidence that Newton was born in 1642, the same year that Galileo died. His birth on Christmas Day by the reckoning of the Julian calendar than in use was uneventful. Newton was a sickly child. He seemed destined to become a farmer as his father had been. Yet his inept management of the family farm at Lincolnshire did little to suggest his enormous scholarly promise; it is unlikely that the daily routine of farm life provided many opportunities for him to demonstrate his intellectual prowess. He was saved from what probably would have been a disastrous farming career by an uncle who was connected with Trinity College at Cambridge. Newton's uncle recognized something of his scholarly potential and secured him a post as a "subsizar," a term for a student obliged to work his way through the university. Newton's student career at Cambridge was not brilliant, but he did receive his degree in 1665, just when the plague made one of its periodic reappearances in London. Newton returned to the family farm at Lincolnshire to avoid the epidemic. In a single year, he (1) discovered the binomial theorem; (2) discovered the basic principles of the differential and integral calculus; (3) worked out the theory of gravitation; and (4) discovered the spectrum, his term for the rainbow of colors resulting when white light is passed through a prism. One would be hard-pressed to imagine any other person who could claim to have made so great a contribution to science in the course of one year. Certainly Newton's achievements at Lincolnshire continued to shape the development of science for the next two centuries. By the time Newton returned to Cambridge in 1667 he was famous throughout the scientific world. In 1669, at the age of twenty-seven, he was appointed to the prestigious Lucasian chair of mathematics at Cambridge. He remained at the university for thirty years. Although he spent most of his time experimenting and thinking, he did give occasional, badly attended lectures. He was elected to the Royal Society in 1672. His contributions to science during his entire Cambridge career were such that for at least two centuries "Newtonian physics" and "classical physics" were synonymous. In the potentially infinite realm of astrophysics and the virtually infinitesimal realm of subatomic particle physics, new physical theories have now eclipsed the usefulness of Newton's work. Yet Newtonian physics is still largely applicable to the world as we know it. Many of his discoveries and inventions—the reflecting telescope, for example—were so important that any single one of them would have sufficed to establish a substantial scientific reputation. It was not until 1685 that Newton settled down to write his masterpiece, spurred on by his loyal and generous friend Edmund Halley. To ensure its international acceptance he wrote it in Latin, entitling it *Philosophiae naturalis principia mathematica*. It was published, largely at Halley's personal expense, in 1687; an English edition, *Mathematical Principles of Natural Philosophy,* did not appear until 1729, two years after Newton's death. A case could be made for considering Newton's mind to be the most powerful intellect the world has known. There is no point in such speculation, however, just as there is no point in passing out first, second, third and consolation prizes to Bach, Beethoven, Handel and Mozart. What is interesting, if not vitally important, is to understand that this towering genius was pretty much of an oddball. His greatest interests were in alchemy and the occult. Newton also feared exposing his thoughts, beliefs, and discoveries to the world, and preferred the solitude of his own company. Fortunately for Newton, he was absorbed in his work with an incredible intensity of concentration. Otherwise, he was tormented by religious problems. Newton was a Unitarian, not a Trinitarian; he believed in God and found no need to accept the divinity of Jesus Christ or even the existence of the Holy Ghost. Such religious views were not always tolerated in Newton's day; as late as 1696, when Newton was fifty-four, a man was hanged for admitting to disbelief in the Trinity.As an adolescent, Newton showed interest in a neighbor girl in Lincolnshire but later dropped her as something of a hindrance to his career. He was never to have romantic dealings with women again. We have little understanding of Newton's sexual nature; it is likely Newton was equally ignorant about the subject. He was cantankerous and suspicious. His quarrels with other scientists—especially Robert Hooke—were notorious for their bitterness and acrimony. A particularly unpleasant controversy began with the independent and almost simultaneous discovery of the calculus by Newton and Leibniz. For several years afterward the question of priority seemed to make little difference, although the British continued to use Newton's symbols in preference to Leibniz's. As the fame of the two men grew and the usefulness of the calculus became more apparent, however, many people started to worry about whether this splendid mathematical system had been invented by an Englishmen or a German. Newton is suspected of having encouraged his supporters in this speculation. Leaving aside the genius of Sir Isaac Newton, as he became known after being knighted by Queen Anne in 1705, it is perhaps better to view the man, his virtues, his defects, through the eyes of his only competitor. Albert Einstein, in his *Essays in Science,* wrote: "Newton himself was better aware of the weaknesses inherent in his intellectual edifice than the generations which followed him. This fact has always aroused my admiration."* The wrote a book. A curious thing about this book is that despite Newton's invention of an entire mathematics as an instrument for his thought, he returned to a virtually Greek, or Euclidean, format when he finally started to systematize his own theories. The main text of the work is divided into three books: I. The Motion of Bodies; II. The Motion of Bodies (in resisting mediums); and III. The System of the World (in mathematical treatment). In addition to his preface, the three books are also preceded by eight Definitions and a *Scholium* that successively treats (1) the flow of "absolute, true, and mathematical time"; (2) the immutability of "absolute space, in its own nature without relation to anything external," constrasting it with "relative space"; (3) place as a part of space; and (4) "absolute motion as the translation [i.e. movement] of a body from one absolute place into another." Newton wrote an additional section entitled "Axioms, or Laws of Motion," which consists of his three famous laws: (1) that every body continues in a state of rest or uniform motion unless its state is changed by forces impressed upon it (what we now refer to as the principle of inertia); (2) that the rate of change in the motion of a body is inversely proportional to its mass and is proportional to the amount of force applied and in the direction in which the force is applied (more commonly known as the principle of acceleration); and (3) that for every action there is an equal and opposite reaction. Newton added six corollaries to the third law as well as an extended *scholium*. Newton also discovered his celebrated formula for gravitation utilizing a constant *G*, the mathematical value of which was determined in 1798 by Henry Cavendish. Newton's belief in absolute time and space was in some ways a remnant of Aristotle's physics. Newton was aware that space and time could not be considered as having completely separate meanings and existences because there was no way to indicate meaningfully a change in time at any point in space. Not only would differently moving observers give one point in space different coordinates but they would not even be able to agree on the identity of points in space. Euclid's geometry is fine for a slate or a sandy beach where points stay put at least some of the time, but it is of no use in outer space where there is no central reference point against which position coordinates can be compared. The absence of an absolute reference system also renders Euclidean geometry useless for systems with relative motion. The points of space imagined by Euclid simply do not exist in nature. Space must be augmented with time and change before it makes any sense in a physics context.Newton was probably aware of the logical flaws in his absolute space, but he used theological arguments to cover up the shortcomings of his physics. He announced that his absolute space was simply the space relative to God. This made it risky to deny the existence of absolute space, since Newton could accuse his critics of denying the existence of God. Earlier devout physicists had resolved this dilemma by declaring that God transcends the mundane frame of space and time.To satisfy Galileo's relativity today even for Newtonian mechanics we take as basic and absolute not the separate time and space of Newton but the concept defined when we give to a physical entity both an instant of relative time and point of relative space together. This concept is called an *event.* Thus a particle at a given point at a given time is an event. Some physicists prefer the more descriptive term *space time point* to the term *event.* What we call space-time (or time-space) in relativity consists of the totality of all space time points. In Newton's mechanics, space by itself is not an objective reality; each observer has his or her own space depending on the observer's motion. The objective reality in Newton's mechanics is not space but time space. In other words, time by itself is an absolute of Galileo's relativity even though space by itself is not. This conclusion predicts that even though two observers in relative motion cannot agree on what will be the same place tomorrow, they can agree on what is the same time. Newton did not leave us a theory of the Universe providing a systematic way to predict the results of all experiments. He did leave us a theory of the planetary orbits of the solar system and a doctrine on which to model the rest of the Universe. Yet Newton was aware of the limits of his world-system; he refused to model gravity, for example, giving his famous disclaimer *"Non fingo hypotheses."* (I do not frame hypotheses.) His success lay in explaining the effect of the gravitational force; he did not seek to uncover its nature.

Neumann, John von. (1903-57), Hungarian-born American mathematician, who taught at Berlin and Hamburg, then entered the US in 1930, becoming a member of the Institute for Advanced Study at Princeton in 1933. He is best known for the founding of Game Theory, but his enormously wide range of work included mathematical economics, computer science, quantum theory, set theory, group theory, operational calculus, probability, mathematical logic and the foundations of mathematics.

Neyman, Jerzy (b. Bendery, Moldavia. Russia, 16.4.1894; d. Berkeley, Calif., USA, 5.8.1981). Polish statistician. Until 1938, when he emigrated to the USA, he had worked in Poland, though making academic visits to France and England. Between 1928 and 1933, with E.S. Pearson, he produced the Neyman Pearson system of hypothesis testing, a set of criteria for maximizing efficiency in the design of test which is the foundation of modern quality control. Faulty items found in an inspected sample may be ineradicable errors from an acceptable situation, or a signal to close down and reset a machine. The theory looks at the cost and risk of stopping a good machine, then at that of running a bad one, to obtain the best criterion and sample size. Many statistical problems are like this. In 1934 Neyman tackled the problems of using random samples in human populations, drawing on experience with the Polish census — considerations of finance, and of rare but important subgroups, mean that real samples are seldom simple. He proposed a general principle: to take linear combinations of data items and balance them first to eliminate bias and then to minimize variance; this gives rules for deciding how intensively to sample, and where. In 1937 he formulated the 'classical' theory of confidence intervals, rejecting postulated values of an underlying parameter if they would imply that the probability of the actual observations was less than a target value. In the long run, the proportion of occasions on which a true value has been rejected will then be less than the target probability.

Nicholas of Cusa (b. 1401; Kues (Cusa), now in West Germany; d. Aug. 11, 1464; Todi, Italy), German cardinal, mathematician, and philosopher. The son of a prosperous fisherman, Nicholas studied at Heidelberg and Padua where he obtained an LLD in 1423. He rose through the hierarchy to become a cardinal in 1448 and a figure of some importance in the Church. His most lasting work was his *De doctaignorantia* (1440; On Learned Ignorance) in which he argued against the possibility of definitive knowledge. "Reason stands in the same relation to the truth as the polygon to a circle," he declared in a striking image. "The more vertices it has, the more it resembles a circle ... yet never becomes a circle." Using metaphysical rather than astronomical arguments he was thus led to deny the standard picture of the universe as bounded by spheres with the Earth at the center. In its place he proposed a universe in which nothing is fixed, with no center or circumference yet not infinite, with everything in motion and a complexity beyond our understanding. He also proposed that the Earth revolved on its axis and around the Sun, and that the stars were other suns. Not surprisingly, such an unorthodox construction had no impact on a conventional thinker like Nicolaus Copernicus. It did; however, help to overthrow in later minds the traditional idea of a 'closed world' and prepare the way for the 'infinite universe' of the 17th century.In other areas of science, Nicholas is also said to have constructed spectacles for the nearsighted (with concave lenses) and to have thought that plants drew nourishment from the air.

Nicolas Sadi Carnot. Nicolas Léonard Sadi Carnot was born in 1796 in Paris, France, into a family made up almost entirely of politicians. His father was Lazare Carnot, the famous engineer; he tutored young Sadi in mathematics, physics, natural sciences, languages, and music. Sadi then studied mechanics, geology, and chemistry at the École Polytechnique before volunteering for military duty. Trained as a military engineer, he was serving in the army at the time of Napoleon's downfall. His father was exiled because of his role in building Napoleon's military organization, and Carnot's own military career was jeopardized by the change in Napoleon's fortunes. But Carnot stayed in the army, and in 1824, he published his sole scientific paper, Réflexions *sur la puissance motrice du feu (On the Motive Power of Heat),* in which he formulated a definition of work as "weight lifted through height." Carnot's interest in heat was thoroughly practical; he needed to know exactly how much work could be performed by a machine operated by heat. The original steam engine invented by James Watt was of extremely low efficiency;

in Carnot's France machines operated at only about six percent efficiency, with some ninety-four percent of the burning fuel being wasted. Carnot developed a celebrated equation, according to which the efficiency of an engine, that is, the maximum fraction of the heat operating a machine that can be converted into work, is the difference between the absolute temperature of the hottest part of the system (steam) and that of the coldest part (coolant) divided by the former. Carnot showed that the laws of thermodynamics should be formulated in terms of cycles or processes in which the system under study, usually an engine, returns to its original state, having effected some desired change in its surroundings. Carnot was among the earliest to distinguish between reversible and irreversible engines. He used a water wheel as a metaphor for all engines, reasoning that it accepts water at one level and delivers it to another level, thereby doing work. The work represents heat and the different levels represent the two different temperatures. This metaphor ultimately led to what we now call the first two laws of thermodynamics. Carnot uses the principles of thermodynamics to prove that of all engines drawing their power from a transfer of heat between two bodies, the most efficient engine is that which operates reversibly — that is, infinitely slowly. Although his work was flawed by his use of the caloric theory of heat, his results, and the second law of thermodynamics, survive today. What we now call a Carnot cycle consists of a cycle of reversible operations in which isotherms (processes without change in temperature) and adiabats (processes without flow of heat) alternate with each other. In short, of all the engines running between the same two temperatures, none is more efficient than the

Carnot engine. Carnot's work remained unknown until about fifteen years after his death in 1832, when it was brought to the attention of the scientific world by Lord Kelvin.

Noether, Amalie Emmy (b. Erlangen, Bavaria, Germany, 23.3.1882; d. Bryn Mawr, Pa, USA, 14.4.1935). German mathematician. Daughter of the mathematician Max Noether and sister of the physicist Fritz Noether, Emmy Noether pursued a distinguished career as a mathematician herself. Her lectures at Göttingen, heard by such men as Artin and van der Waerden, strongly influenced the development of abstract algebra in 20c mathematics. Her own research work continued the development of the theories of rings and ideals, largely in the style of the 19c German mathematician Richard Dedekind, of whose collected works she was a coeditor in the 1930s. She also contributed to the theory of hypercomplex numbers (a generalization of complex numbers). As a woman, Noether was not able to become a full professor at Göttingen; as a Jewess, she even lost her extraordinary professorship in 1933, and she left for the USA where she obtained a post at Bryn Mawr College. She died there, suddenly, two years later. A. Dick, *Emmy Noether 1882-1935* (Boston, Mass., 1981).

Nϕrlund, Niels Erik. (b. Oct. 16, 1885; Stagelse, Denmark) Danish Mathematician and geodesist. Nϕrlund studied at Soro High School and then worked as anassistant at the Copenhagen Observatory. His work there was to influence many of his subsequent scientific interests. After obtaining his doctorate in 1910 from the University of Copenhagen, Nϕrlund took up a post in Sweden as professor of mathematics at Lund University. He remained there until 1922 when he returned to Denmark to become professor of mathematics at Copenhagen University, a post that he held until his retirement in 1956. He held many official posts, among them director of the Royal Geodesic Institute of Denmark (1923-55) and editor of the journal *Acta Mathematica.* Nϕrlund's interest in geodesy dates back to his time at the Copenhagen Observatory and his publications include works on the mapping of both Denmark and Iceland. He organized a new triangulation of Denmark from 1923, accompanied by gravity measurements and astronomical determinations of longitude, and also carried out a partial triangulation of Greenland. Among his more purely mathematical interests were differential equations and analysis.

Norman R. Campbell. Norman R. Campbell, the noted physicist and science philosopher, asserted on more than one occasion that physics is not dependent on mathematical conceptions but is derived purely from experiments. But Campbell was forced to admit that mathematical conceptions play an important role in physics. In determining whether numerical laws involve mathematical conceptions, Campbell reasoned that "the facts expressed by numerical laws can be expressed as laws, and that these laws can be used to define derived magnitudes and for any other purpose for which any other laws are used, without any introduction of mathematical conception." He went so far as to suggest that these laws could be expressed as relations between numerals without introducing such conceptions. However, Campbell admitted that "if we inquire how we obtain the numerals by which numerical laws are expressed, then we are bound to introduce the mathematical conception of Number." Campbell believed that mathematical conceptions are needed only to explain and understand numerical laws. Numerical laws test physical theories, which brings mathematics into the fold of physics. Such physical theories usually involve the mathematical conceptions of limits, continuity, derivatives, and integrals. Although he was particularly aware of the utility of mathematics in the sciences, "Campbell viewed as mistaken the argument that mathematical theories are less likely to lead the scientist into error, on the grounds that they are, as Mach put it, 'purely phenomenal,' since considerations of simplicity of the kind which Campbell identified as characteristic of mathematical theories cannot be considered as determined solely by 'phenomena.' Indeed, in an important sense, mechanical theories are more 'purely phenomenal' since the propositions of their hypotheses are analogous to true observational laws."* Campbell was careful to distinguish between mathematical and nonmathematical theories, though he recognized that whether a theory is ultimately accepted as mathematical or not often depends on research by other physicists. He was also doubtless aware that the "explanatory power of a theory is dependent on the 'intrinsic interest' which it may have for a particular scientist or school of scientists."** For example, Campbell did not consider Maxwell's theory intrinsically mathematical until Hertz's experiments gave it the requisite mathematical foundation by proving the existence of the displacement current. Campbell believed that the application of mathematics to a physical theory introduces a symmetrical ordering of that theory enhancing both its simplicity and elegance. "As is characteristic of such an experimentalist, Campbell's inclinations were toward a quantitative view of his science. Indeed, for him physics was the 'science of measurement,' and he devoted much space in his texts to considerations of the nature of physical measurement and its relation to purer mathematics. Once again his firm commitment to the role of theories appeared in his insistence that 'no new measurable quantity has ever been introduced into physics except as a result of the suggestions of some theory.' "

O

Omar Khayyam. (b. c. 1048; Nishapur, now in Iran; d. c. 1122; Nishapur). Persian astronomer, mathematician, and poet. Omar Khayyam produced a work on algebra that was used as a textbook in Persia until this century. He gave a rule for solving quadratic equations, he could solve special cases of the cubic, and — in a last work — seemed to have some inkling of the binomial theorem. He also worked on the reform of the Persian calendar, which was basically the Egyptian one of 365 days, introducing a sixth epagomenal (extra) day and obtaining an accurate estimate of the tropical year.

Oughtred, William. (b. Mar. 5, 1575; Eton, England; d. June 30, 1660; Albury, England). English mathematician. Oughtred was educated at Eton College, where his father taught writing, and at Cambridge University, England. He was ordained a priest in 1603 and eventually became rector of Albury. Despite his clerical post he found time to work on mathematics and he produced what was to become a very famous book on mathematics, the Clavis mathematicae (1631; The Key to Mathematics). This work dealt with arithmetic and algebra, and it is of historical importance because Oughtred managed to put into it more or less everything that was known at that time in those areas of mathematics. It rapidly became an influential and widely used textbook and held in high regard by mathematicians of the stature of Isaac Newton and John Wallis, himself a pupil of Oughtred. A number of mathematical symbols that are still used were first introduced by Oughtred. Among these were the sign 'X' for multiplication, and the 'sin' and 'cos' notation of trigonometrical functions. Oughtred also invented the earliest form of the slide rule in 1622 but only published this discovery in 1632, and as a result became embroiled in a violent dispute with one of his former students, Richard Delamain, who had made the same invention independently. Oughtred's religious views were conservative and he was a staunch supporter of the Royalist party, but during the time of Cromwell and the Commonwealth he was able to retain his post as vicar. He lived just long enough to see Charles II installed on the throne.

P

Pappus of Alexandria. (fl. 320) Greek mathematician. Pappus was the last notable Greek mathematician and is chiefly remembered because his writings contain reports of the work of many earlier Greek mathematicians that would otherwise be lost. His chief work, *Synagoge* (c. 340; Collection), consisted of eight books of which the first and part of the second are now lost. It was intended as a guide to the whole of Greek mathematics and this is what makes it such a significant historical source. Among the mathematicians whose work Pappus expounds are Euclid, Apollonius of Perga, Aristaeus, and Eratosthenes. Pappus did contribute some original work, however, notably in projective geometry. As with many Greek mathematicians Pappus was as interested in astronomy as in pure mathematics and his other work included comments on Ptolemy's astronomical system contained in the *Almagest*.

Pascal, Blaise. (b. June 19, 1623; Clermont-Ferrand, France; d. Aug. 19, 1662; Paris). French mathematician, physicist, and religious philosopher. Pascal's father was respected mathematician and a local administrator. Early in life Pascal displayed evidence that he was an infant prodigy and apparently discovered Euclid's first 23 theorems for himself at the age of 11. While only 17 he published an essay on mathematics that René Descartes refused to acknowledge as being the work of a youth. Pascal produced (1642-44) a calculating device to aid his father in his local administration; this was in effect the first digital calculator.Pascal conducted important work in experimental physics, in particular in the study of atmospheric pressure. He tested the theories of Evangelista Torricelli (who discovered the principle of the barometer) by using mercury barometers to measure air pressure in Paris and, with the help of his brother-in-law , on the summit of the Puy de Dôme (1646). He found that the height of the column of mercury did indeed fall with increasing altitude. From these studies Pascal invented the hydraulic press and the syringe and formulated his law that pressure applied to a confined liquid is transmitted through the liquid in all directions regardless of the area to which the pressure is applied. He published his work on vacuums in 1647. Pascal corresponded with a contemporary mathematician, Pierre de Fermat, and together they founded the mathematical theory of probability. Pascal had been converted to Jansenism in 1646 and religion became increasingly dominant in his life, culminating in the religious revelation he experienced on the night of November 23, 1654. Following this he entered the Jansenist retreat at Port-Royal (1655) and devoted himself to religious studies from then on.

Paul A. M. Dirac. Paul Dirac was acknowledged during his lifetime as one of the most creative theoretical physicists and mathematician of the first half of this century. Not only did he derive a general formalism for quantum mechanics in 1926 that incorporated Heisenberg's matrix mechanics and Schrödinger's wave mechanics, but he formulated a relativistic theory to describe the nature of the electron, improving upon the nonrelativistic wave equations of the electron developed by Schrödinger. Dirac's equations not only predicted the correct energy levels of the hydrogen atom but also showed that some of these energy levels could be divided into two parts, indicating that electrons have spin. Dirac turned his attention to cosmic numbers and, in 1937 a paper entitled *The Cosmological Constants* which examined some of the large number coincidences—that is, the numerical relationships that appear to exist between some of the constants found in nature. Dirac found that 10^{40} is particularly prominent as it represents the ratio of the force of electrostatic attraction between a proton and an electron to the force of gravitational attraction between the two particles. The known radius of the Universe is also about 10^{40} times the radius of an electron. Finally, 10^{40} represents the square root of 10^{80} —the number of particles in the cosmos. Like Eddington, Dirac saw that there might be some fundamental significance in this particular cosmic number. Yet Dirac approached the question more cautiously, hypothesizing a model of the Universe that relies extensively on the number 10^{40} and arguing that there is a subtle connection between the force ratio and the radius ratio mentioned above. However, his model was not particularly helpful in establishing the nature of this connection although he stated that due to the continuing expansion of the radius of the Universe, the force ratio, which is dependent on the gravitational variable, may decrease with time.

Peano, Giuseppe (b. Aug. 27, 1858; Spinetta, near Cuneo, Italy; d. Apr. 20, 1932; Turin, Italy). Italian mathematician and logician. Peano studied at the University of Turin and was an assistant there from 1880. He became extraordinary professor of infinitesimal calculus in 1890 and was full professor from 1895 until his death. He was also professor of the military academy in Turin from 1886 to 1901. Peano began his mathematical career as an analyst and, like Richard Dedekind before him, his interest in philosophical and logical matters was awakened by the lack of rigor in some presentations of the subject. Peano was particularly keen to avoid all illegitimate reliance on intuition in analysis. His discovery in 1890 of a curve that was continuous but filled space went against intuition. A similar discovery was Karl Weierstrass's famous function that was everywhere continuous but nowhere differentiable. As with Weierstrass's function, Peano's curve shows that the concept of a continuous function cannot be identified with that of a graph. His interest in rigorous and logical presentation of mathematics led Peano naturally to an interest in the mathematical development of logic. In this field he was one of the great pioneers along with George Boole, Gottlob Frege, and Bertrand Russell. Peano's achievement was twofold. First he devised a clear and efficient notation for mathematical logic that is in many ways much more satisfactory than either Frege's or Russell's. Secondly he showed how arithmetic can be derived from a purely logical basis. To do this Peano formulated a set of axioms that captured precisely the logical concept of the natural numbers. These axioms are now known as the *Peano axioms* (or postulates), despite the fact, which Peano himself pointed out, that they had first been formulated by Richard Dedekind. Peano also did notable work in geometry and on the error terms in numerical calculation. Among his extra-mathematical interests he was a keen propagandist for a proposed international language, Interlingua, which he had developed from Volapük.

Pearson, Egon Sharpe. (b. London, UK, 11.8.1895; d. Midhurst, W. Sussex, 12.6.1980). British statistician. Son of Karl Pearson, Egon Pearson held the chair of statistics at University College London from 1935 to 1960. He collaborated with Neyman in producing the classic series of papers on statistical inference, developing such concepts as the likelihood ratio test of an hypothesis. This states that an hypothesis is not invalidated because it makes observed events improbable; there must be a realistic alternative hypothesis that does better. If a coin is tossed 50 times, and gives 'heads' 50 times, it may still be fair. If we allow that a bias of 53 percent to heads is physically possible, then that hypothesis is marginally better. Only if we allow for a bias as large as 58 percent is fairness inconceivable. Unfortunately, the problem of deciding which hypotheses to consider is often intractable.

Pearson, Karl. (b. London, UK, 27.3.1857; d. Coldharbour, Surrey, 27.4.1936). British statistician. Trained in the law, he turned to mathematics and became professor of applied mathematics at University College London. He was drawn to the study of genetics following Galton and became Galton professor of eugenics. He clarified Galton's work on correlation, and established the technique of multiple regression. If one quantity, such as the width of a crab, has its relationship to another masked by nuisance variables, or is not to be predicted by one variable alone but by two or three, then all the variables are put into a linear equation, and the relationship expressed by a partial or a multiple correlation. In 1900 Pearson produced the chi-squared 'measure of fit', perhaps the most used of individual statistics. When a fixed number of observations are counted into cells, the deviations from the expected frequencies are not independent. Having found a surplus, a corresponding deficit contains little new information. Pearson's statistic is a balanced summary, and allows an estimate of whether so large a discrepancy could occur by chance alone. The Pearson distributions were designed to allow measured departures from the normal curve in symmetry or in the proportion of height to spread. They have proved to be ubiquitous, especially in showing the effects on derived statistics of the possible choices of samples. Pearson wrote an important treatment of *The Life of Galton* (London, 1914-30). E. S. Pearson, *Karl Pearson* (Cambridge, 1938).

Penney, William George, Baron Penney of East Hendred (b. June 24, 1909; Gibraltar). British mathematician. Penney was educated at London University, the University of Wisconsin, and Cambridge University, where he obtained his PhD in 1935. He taught mathematics at Imperial College, London, from 1936 until he became involved in the development of nuclear weapons in the war; he served at Los Alamos and witnessed the dropping of the Nagasaki bomb. After the war he worked on the development of the British bomb at the Ministry of Supply from 1946 until 1952, when he was appointed director of the Atomic Weapons Research Establishment at Aldermaston. Penney left Aldermaston in 1959 to join the Board of the Atomic Energy Authority on which he served as its chairman from 1964 until 1967, when he returned to academic life as rector of Imperial College, a post he held until his retirement in 1973. Penney was the key figure in Britain's success in exploding their first atomic bomb at Monte Bello in the

Pacific in 1952. He also worked on the production of the British H-bomb, first tested on Christmas Island in 1957.

Penrose, Roger (b. Aug. 8, 1931; Colchester England). British mathematician and theoretical physicist. Penrose, the son of the geneticist Lionel Penrose, graduated from University College, London, and obtained his PhD in 1957 from Cambridge University. After holding various lecturing and research posts in London, Cambridge, and in America at Princeton, Syracuse, and Texas, Penrose was appointed professor of applied mathematics at Birkbeck College, London, in 1966. In 1973 he was elected Rouse Ball Professor of Mathematics at Oxford. Penrose has done much to elucidate the fundamental properties of black holes. These result from the total gravitational collapse of large stars that shrink to such a small volume that not even a light signal can escape from them. There is thus a boundary around a black hole inside which all information about the black hole is trapped; this is known as its 'event horizon'. With Stephen Hawking, Penrose proved a theorem of Einstein's general relativity asserting that at the center of a black hole there must evolve a 'space-time singularity' of zero volume and infinite density where the present laws of physics break down. He went on to propose his hypothesis of 'cosmic censorship': such singularities cannot be 'naked' —they must posses an event horizon. The effect of this would be to conceal and isolate the singularity with its indifference to the laws of physics. Despite this Penrose went on the 1969 to describe a mechanism for the extraction of energy from a Kerr black hole, an uncharged rotating body first described by Roy Kerr in 1963. Such bodies are surrounded by an ergosphere within which it is impossible for an object to be at rest. If Penrose demonstrated, a body fell into this area it would split into two particles; one would fall into the hole and the other would escape with more mass-energy than the initial particle. In this way rotational energy of the black hole is transferred to the particle outside the hole. Penrose has also done much to develop the mathematics needed to unite general relativity (which deals with the gravitational interaction of matter) and quantum mechanics (which describes all other interactions).

Philip Jourdain. Philip Jourdain was a prominent logician, philosopher, and mathematical historian who brought to each of these subjects a remarkably creative and penetrating intelligence and made substantial contributions that greatly influenced the development of mathematical logic and the natural sciences. The son of a vicar, Jourdain was educated at Cheltenham College and Cambridge University. Yet his childhood was overshadowed by early signs of the hereditary paralytic disease that would kill him in 1919 before his fortieth birthday. He traveled to Heidelberg to seek medical help. Although the treatments did nothing to alleviate his suffering, Jourdain began to study mathematics during his hospital stay. He had difficulty walking and even holding a pencil steady, but Jourdain approached his newfound interest with great enthusiasm and began to write a series of papers that established his reputation. Some of these papers were about subjects as diverse as Lagrange's use of differential equations and conceptual problems of mathematical physics. He contributed a valuable series of papers on transfinite numbers to the *Archiv der Mathematik und Physik* and also published another group of papers on mathematical logic and the principles of mathematics in the *Quarterly Journal of Mathematics.* He also worked as an editor for George Sarton's *Isis* as well as the *Monist* and wrote articles about a number of prominent scientists and mathematicians ranging from Isaac Newton to Henri Poincaré. Jourdain is perhaps best remembered for a little book called *The Nature of Mathematics* that explores the development of a number of mathematical methods, especially what he called "analytical methods and certain examinations of principles." The following selection illustrates how the development of increasingly rigorous approaches to physics, especially dynamics, necessitated a correspondingly greater reliance on mathematics. In Jourdain's opinion, this was not an unanticipated development since he regarded numbers as the way to quantify ideas about the workings of the Universe. Ever the logician, Jourdain examines not only the dependence of natural science on mathematics but also the reliance of mathematics on its own logical consistency. Jourdain's book also studies the relation between mathematics and logic.

Pierre Duhem. Pierre Duhem demonstrated talent in mathematics and physics and entered the Ecole Normale Supérieure on the rud d'Ulm in Paris in 1881 at the age of twenty. A brilliant student, he was especially interested in thermodynamics and its applications. His reflections on the analogy between Lagrange's analytical mechanics and thermodynamics led him to the notion of thermodynamic potential and the publication of a book on that subject in 1886. He began his teaching career as a lecturer in the Faculty of Sciences of Lille University where he taught acoustics, elasticity, and hydrodynamics. He became a full professor at the age of thirty-two in the

Faculty of Sciences of Bordeaux University where he remained until his death. During his scientific career, which was particularly notable for his research in thermodynamics, Duhem developed an extensive knowledge of the history of mathematical physics and published a number of papers containing his views about the subject. Duhem's mind was precise and orderly. He felt that theories should be constructed by axiomatic methods anchored in exact postulates to derive undisputable conclusions by logical reasoning. He preferred the formal arguments of the energeticists to the unclear models of the atomists; Duhem did not desire to follow Maxwell and Boltzmann in their construction of a kinetic theory of matter permitting less abstract interpretations of particular conceptions of thermodynamics. According to Louis de Broglie, Duhem "sifted out all the fundamental notions admitted by thermodynamics; for example, he gave a purely mathematical definition of the quantity of heat and thus deprived it of any physical intuitive meaning in order to avoid any begging of the question." Duhem devoted much of his attention to hydrodynamics and the theory of elasticity. He also studied the propagation of waves in fluids and did research in electromagnetism, although his preference for the theories of Helmholtz instead of Maxwell in the latter category ensured the obsolescence of his own work in electromagnetism. He also could not accept Lorentz's theory of the electron or the discoveries in the emerging field of atomic physics. A student of the history of science, Duhem published a three-volume work which suggested that the revival of mechanics, astronomy, and physics during the Renaissance was due in part to the vital role played by scholars in the medieval universities. Leonardo da Vinci was the hero of this work who integrated the disparate findings of his predecessors in mechanics and astronomy and paved the way for Galileo's development of modern mechanics. Duhem also completed eight volumes of a monumental history of ideas about the cosmos from antiquity to his own era. Duhem sought to separate physics from metaphysics by limiting the former to "a system of mathematical propositions whose aim is to represent as simply, as completely, and as exactly as possible a whole group of experimental laws." Metaphysics, by contrast, was to be rooted in religious revelation. As a result, Duhem viewed physical theory as a method of classification of physical phenomena. Only mathematical, specifically algebraic, relations established by well-founded theories were acceptable to Duhem. Although he was criticized as being too narrowminded due to his reliance on mathematical relations in physical theories, his approach confirmed the importance of mathematics to modern physics.

Pierre Simon De Laplace. Pierre Simon De Laplace lived from 1749 to 1827, a period of internal upheavals in France, yet he managed to pursue science with a single-minded devotion that won him many honors and established him as the first great post-Newtonian astronomical theorist. When only eighteen years old, he arrived in Paris from the provinces armed with a letter of introduction to Jean le Rond D'Alembert, who declined to receive him. Undaunted, young Laplace responded to this discourtesy by sending D'Alembert an essay on mechanics of such brilliance that the distinguished scientific editor of Diderot's Encylopédie was quite won over and used his influence to get Laplace a job teaching mathematics. Soon Laplace was working on chemistry with no less a master than Lavoisier. Their experiments on the determination of the specific heat of various substances led to a seminal discovery that became the basis of thermochemisty. But Laplace's true interest was astronomy. he started studying perturbations of the bodies making up the solar system. He made many valuable contributions to the understanding of variations in the gravitational interplay of planets. His conclusions on gravitational theory are set forth in his five-volume La Mécanique céleste (Celestial Mechanics), published in parts over a twenty-six-year period from 1799 to 1825. His political mechanics proved to have been studied with equal devotion, and in a period when many a fine head was being separated from its owner's body (*se faire raccourcir*—"to get shortened"—was the slang term for the operation), Laplace managed to get along well with the various regimes which followed one another. Napoleon made Laplace his Minister of Interior Affairs (or attorney general), but the distinguished scientist proved no great talent as a cabinet minister and so he was given a sinecure as an appointed (i.e. nonelective) senator. After the fall of Napoleon, his successor Louis XVIII threw out most of Bonaparte's appointees. Yet Laplace stayed on and was granted the title of marquis (an honor which would have cost him his head some decades earlier). Laplace used Newton's second law of motion together with Newton's law of gravity to organize the subtlest trepidations of the solar system into an understandable whole. Yet he also gave us one of the most important tools for replacing this mechanical world system by a field theory: the concept of gravitational potential. The gravitational potential of a body such as a planet is a variable whose value at each point in space is the work required to bring a small probe from that point to a standard surface such as infinity. It is called potential because that work is in effect stored in the system, ready to be converted into kinetic energy or work by letting the probe fall from the infinite surface toward the point. The equation this potential obeys (which is called Poisson's equation) says that outside a point mass the gravitational potential at any point is proportional to the mass divided by its distance from the point. By relating values of the quantity at one point to values in the infinitesimal neighborhood of that point, Laplace provided a model for all future field theories. He also reduced the three components of force of Newton's second law to the space rate of change (gradient) of potential energy so that it could be reexpressed as follows: The rate of change of the momentum at a point is equal to the negative of the gradient of the potential at that point. The word "gradient" means rate of change with respect to position. As any position in space is described by three coordinates, three components of force result from one potential energy. This simplification may be clarified by an analogy. When a ball is placed on a hillside it experiences a force tending to make it roll downhill. Newton's doctrine would have us search for the force on the ball. The force has a component along the North-South direction and along the East-West direction on the hillside; both are required by Newton's law of motion. Laplace points out that it is enough to know the height of all hill at each point. Knowledge of the profile of the hill about a point enables us to construct the gradient of the hill at that point. Once we obtain the gradient we can compute the force by multiplying the magnitude of the gradient by the mass of the ball. As a result, two force components emerge from one height component. The three components of the force in three-dimensional space similarly arise from the gradient of one potential. The gradient can be taken in any direction to give the force in that direction. In his most famous aphorism. Laplace boasted that a perfect intellect armed with Newton's second law and knowledge of the present state of the world (where "state" means the positions and velocities of all bodies at one instant) could compute the entire future history of the Universe. Perhaps Laplace only intended to express the fact that the initial state of the world determines all future states, without suggesting that there is access to such knowledge. But Laplace's statement is paradoxical. To predict the history of the entire Universe, the perfect intellect must also, of necessity, predict its own behavior, since it is a part of the Universe. But to predict its behavior it must contain within itself an exact representation of itself. Moreover, the exact representation must itself contain a representation and so on *ad infinitum*. This infinite paradoxical recursion shows us that for Laplace's perfect intellect to do what Laplace says it will do, it must be impotent as well as omniscient; outside the Universe helplessly looking in. Pierre Simon De Laplace, the French astronomer and mathematician, is generally credited with having wondered about the effects a very massive star would have on light although his thoughts were anticipated twelve years before they appeared in the 1796 publication of his monumental *Exposition of the system of the World* by John Mitchell, an English rector and astronomer. Mitchell realized in 1784 that "[a]ll light emitted from [a sufficiently massive] body would be made to return to it, by its own power of gravity." Mitchell assumed that light consists of corpuscles or particles and that they obey Newton's laws of motion. Laplace may have been aware of Mitchell's work but it is just as plausible that his extensive work with Newtonian mechanics caused him simply to carry the concept of gravitational attraction to its ultimate conclusion. Laplace predicted "that the attractive force of a heavenly body could be so large that light could not flow out of it." Yet Laplace's idea never attracted much attention and remained more of a historical curiosity until the first half of the twentieth century when the implications of Einstein's general theory of relativity began to be fully explored. Although black holes remain in some ways as physically inaccessible today as they were when Laplace and Mitchell were alive, we at least have the equations of Einstein's theory to explain mathematically the collapse to a black hole state. Laplace and Mitchell, on the other hand, had only Newton's laws of motion and his *Principia Mathematica*. Laplace did gain much fame by using Newtonian mechanics to formulate his nebular hypothesis, a speculation he offered at the end of a nonmathematical book that he himself did not take too seriously, As the planets had been observed to revolve around the sun in the same direction and almost coincident planes, Laplace suggested that the solar system may have originated in a swirling cloud of gas which later condensed into a star and planets. Unlike the black hole suggestion, Laplace's nebular hypothesis caught the public fancy and remained popular throughout much of the nineteenth century; after a period of dormancy, it was revived in the mid-twentieth century by Carl Friedrich von Weizsäcker. Unlike the massive star hypothesis, the origin of the solar system seemed a much more relevant topic especially when the debate over evolution and the creation of the world began to heat up in the last century.

Plato. Plato was not only one of the most creative thinkers of ancient Greece; he was also one of its most durable philosophers. His influence throughout the Middle Ages and Renaissance, in fields ranging from natural science to politics, was powerful. Even today nonspecialists read him with keen delight. He was born about 428 B.C. of aristocratic Athenian stock. As a young man he had political ambitions, but he was repelled by the excesses of the oligarchy holding power in 404-403 B.C. (the rule of the "Thirty Tyrants"), and especially by the execution in 399 B.C of his teacher Socrates, who was also an old family friend. In 388 B.C. he visited Italy and Sicily, where he enjoyed the friendship of Dionysius the Elder, the ruler of Syracuse, and his brother-in-law, Dion. In 387 B.C. he returned to Athens, where he founded his celebrated Academy. On the death of Dionysius, Dion invited him to direct the education of the new ruler, Dionysius the Younger. This venture and a later visit to the Sicilian court were both unsuccessful, and by 360 B.C. Plato was back at his Academy. He died in 348 or 347 B.C. Plato's Timaeus consists of accounts of contemporary, predominantly Italian, natural sciences ranging from astronomy to chemistry to anatomy. The original views of nature that do appear in Timaeus were shaped by Plato's Theory of Forms, which focused on five regular forms chosen by Plato and their presumed significance in nature. The Theory of Forms portrayed the real world of soil, clouds, and stars as being transcended by a second, superior world containing "concept-objects." In the same way that "flower" or "child" signified an apparent object in this world, so did "truth" or "justice" identify mental images in the transcendent world. In Timaeus, Plato described God as a creative force who fashioned all objects in the Universe after unchanging, geometrically perfect concept objects or abstractions such as cubes and pyramids. For Plato, the perfection and divinity of the Universe lay in its adherence to this succession of geometrical forms. The Theory of Forms also laid the groundwork for later attempts to describe the world in terms of geometrical patterns, although this reverence for perfect forms severely hindered the creativity of later Greek astronomers. Plato's philosophy is presented in the form of dialogues, discussions usually led by Socrates, who questions various groups of other characters. The passages from the *Timaeus* included here provide Plato's conception of the origin of both the Universe and mankind.

Playfair, John (b. Mar. 10, 1748; Benrie, Scotland; d. July 20, 1819; Edinburgh). Scottish mathematician and geologist. Playfair studied at St. Andrews University and became minister of Liff and Benvie in 1773. He was made a professor of mathematics at Edinburgh University in 1785 and, in 1805, professor of natural philosophy. Playfair was a friend of the geologist James Hutton (q.v.) and in his *Illustrations of the Huttonian Theory of the Earth* (1802) he amplified and explained Hutton's uniformitarian ideas. Hutton's own work had been notoriously hard to follow and Playfair brought uniformitarianism to a considerably larger public. He also pioneered the idea that a river carves out its own valley. Although he is better known as a geologist Playfair did make contributions of note to mathematics, in particular to geometry. In 1795 he published his *Elements of Geometry* in which he set out an alternative version of Euclid's fifth postulate, which, given the truth of the other postulates, is equivalent to Euclid's original formulation. This postulate is consequently now known as 'Playfair's axiom' and asserts that for any line (L) and point (P) not on L there is one and only one line L' through P parallel to L.

Plücker Julius (b. June 16, 1801; Elberfeld, now in West Germany; d. May 22, 1868; Bonn). German mathematician and physicist. Plücker studied in Heidelberg, Berlin, and Paris, and became a professor of mathematics at the universities of Halle and Bonn. He became professor of physics at Bonn in 1847. He did important and pioneering work in analytic geometry in which he suggested taking straight lined rather than points as the basic geometrical concept. This idea led him to the celebrated principle of duality, stating the equal validity of certain equivalent theorems. He used this in a geometrical context but it now has far wider applications. In about 1847 Plücker turned from mathematics to physics, studying the discharge of electricity through gases. He was the first to find that cathode rays can be deflected by a magnetic field, thus indicating their electric charge.Later Plücker returned to his mathematical interests and did further work on geometry.

Plumley, J.M. J. M. Plumley is a distinguished professor of Egyptology at England's Cambridge University. In his writings Plumley has stressed that the modern reader is separated from ancient Egypt in time by a hiatus of nineteen centuries and psychologically by an insurmountable gap, since Egyptian thought was based on a set of beliefs of which we know virtually nothing. In spite of the extraordinary advances in the decipherment of hieroglyphic texts, little is known of the grammar and lexicography of the ancient Egyptian language, so that much of the figurative syntax—those similes and metaphors and other manners of speaking that are the soul of a language—is wholly lost on the modern mentality.

A simple illustration of the diversity of the ancient and modern mentalities is that while we think of southern Europe and Egypt as facing each other from opposite shores of the Mediterranean, the Egyptians thought of themselves as facing south, with the sea and the rest of the world behind them, the west lying to the right and the Orient to the left. Vestiges of this ancient orientation remain today as what is Upper Egypt is actually southern Egypt and what is Lower Egypt is the region north of Cairo. The annual flooding of the Nile coincided with the helical rising of the star Sirius. The need to be able to predict the changes of the seasons gave rise to a priesthood which knew enough astronomy to devise a calendar and regulate the irrigation of the fertile crescent. The Egyptians were skilled in hydraulics and civil engineering; their pyramids and temples remain the supreme surviving architectural monuments of the ancient world. Yet little of their technology can be found in the writings that remain. These writings do provide some clues to the abstract theories that grew along the Nile. Egyptian arithmetic, for example, was a formidable tool of trade and taxation. The earliest versions of the cosmologies of Pythagoras, Heraclitus, and Thales may have first appeared in Egypt. Its bustling ancient cities may have also been where Phoenician merchants found the seeds of our own alphabet, an innovation about which no less a figure than Galileo stated: But of all other stupendous inventions, what sublimity of mind must have been his who conceived how to communicate his most secret thoughts to any other person, though very far distant in time or place, speaking with those who are in the Indies, speaking to those who are not yet born, nor shall be this thousand or ten thousand years? And with no greater difficulty than the various arrangements of two dozen little signs upon paper? Let this be the seal of all the admirable iventions of man. The climate of Egypt today, as nineteen centuries ago, is one in which torrid days, with the dry land baking beneath a steely, cloudless blue sky, lead to bitter cold nights in which the stars glitter like shards of ice. The day has the warmth and noise of life, the night the coldness and silence of death. It is little wonder that an important figure in the Egyptian pantheon was that divinity so often found in crossword puzzles, Ra the Sun God.

Poincaré, Jules Henry. (1854 - 1912), Prolific French mathematician and physicist who was Professor of Mathematics and Science at the *Université de Paris,* and made major contributions to virtually all branches of mathematics. He originated the study of Automorphic Functions, was a pioneer of topology, an astronomer, a probability theorist, a philosopher, and a member of the Académie Francaise, and became President of the French Academy of Science. Poincaré studied at the Ecole Polytechnique and the School of Mines. At first he had intended to become an engineer, but fortunately his mathematical interests prevailed and he took his doctorate in 1879 and then taught at the University of Caen. He was professor at the University of Paris from 1881 until his death. In pure mathematics Poincaré originated and wellnigh completed the theory of automorphic functions and in probability theory he introduced the important concept of ergodicity. He was an early pioneer in giving a systematic treatment of topology and made notable contributions to the theory of numbers. Poincaré's views on the philosophy of mathematics were in general very hostile to the 'logicist' ideas of Gottlob Frege and Bertrand Russell. Poincaré stressed the importance of mathematical induction as a nonlogical principle. One of Poincaré's important contributions to applied mathematics consisted in his attacks on the three-body problem and the general n-body problem; i.e. the problem of calculating the motions of n bodies (e.g. planets) under their mutual interactions. Poincaré did not give a complete solution but he was able to make highly significant advances by producing methods by which solutions could be approximated. Among the techniques he developed for this purpose were the theory of asymptotic expansions and integral invariants. Other notable contributions made by Poincaré to applied mathematics include his work on the theory of the states of equilibrium of a rotating fluid mass, and in mathematical astronomy his work on differential equations led to a partial anticipation of Einstein's work on the special theory of relativity. Poincaré was remarkable for the extreme range of his mathematical interests and the speed at which he produced mathematics. He was equally quick in verbal discussion and in composing his own papers. Although he died at relatively early age of 58, he produced almost 500 mathematical papers on subjects ranging from physics and mathematical astronomy to pure mathematics and philosophy. Raymond Poincaré, president of France from 1913 to 1920, was his cousin.

Poincaré, Jules Henri (b. Nancy, Meurthe-et-Moselle, France, 29.4.1854; d. Paris, 17.12.1912). French mathematician and philosopher. Perhaps the most brilliant and prolific mathematician of the second half of the 19c, Poincaré made important contributions to virtually all aspects of pure and applied mathematics, and was also influential as a philosopher of mathematics and of science. Poincaré's chief mathematical interest was differential equations. They led him in the 1880s to important contributions to 'automorphic functions', and later to existence theorems on solutions. Among applications of differential equations, he wrote extensively on the three-body (Sun-Moon-Earth) problem: the 'Poincaré recurrence theorem' is a remarkable result on the recurrence of planetary orbits. He also studies planetary physics, producing important conclusions on possible shapes in which the earth could be in equilibrium. He wrote extensively on heat and thermodynamics, and on probability, where he laid the foundations of ergodic theory. In pure mathematics he introduced a number of notable innovations into topology. Poincaré's philosophy of science is called 'conventionalism'. For him a scientific theory was a convention, of which one demands 'simplicity' and generality of application; one is not concerned with its truth. This view guided his contributions to the foundations of mechanics in the 1900s, where he saw Newton's laws as 'definitions in disguise' and rather opposed Einstein's theory of relativity, although he enunciated the principle of relativity himself. In the philosophy of mathematics his position was similar to Kant's, and he opposed both the formalism of Hilbert and the logicism of Bertrand Russell. He gathered his philosophical writings into popular books (*La science et l'hypothèse*, Paris, 1903; *La valeur de la science,* Paris, 1905; *Science et méthode*, Paris, 1908; all tr. G. B. Halsted in one volume as *The Foundations of Science*, London, 1913). The English translations of these and of *Derniéres Pensées* (Paris, 1913; *Mathematics and Science, Last Essays,* London, 1963) are still in print. In 35 years he produced, on average, a book a year (including editions of his lectures at the Sorbonne) and also about 500 papers. His papers are collected in *Oeuvres* (11 vols,

Paris, 1916-54).

Poisson, Simeon-Denis. (b. June 21, 1781; Pithiviers, France; d. Apr. 25, 1840; Sceaux, France). French mathematician and mathematical physicist. Poisson, the son of a local government administrator, studied at the Ecole Polytechnique, where his teachers included Pierre Simon Laplace and Joseph Lagrange, and later he himself held various teaching posts at the Ecole. His important mathematical work was largely in mathematical physics and he also did a considerable amount of experimental work on heat and sound. In thermodynamics he played an important role in making the whole subject amenable to mathematical treatment by showing how to quantify heat precisely. He is also one of the principal founders of the mathematical theory of elasticity. Poisson is possibly best known for his work on probability, and he was something of a pioneer in applying the techniques of mathematical probability to the social sciences, something that was extremely controversial at the time. The term 'law of large number' was introduced by Poisson in his seminal work, *Recherches sur la probabilité des jugements* (1837; Researches on the Probability of Opinions), in which he put forward his discovery of the *Poisson distribution.* This is the distribution that is a special case of the binomial distribution obtained when the probability of success in a given trial is some constant divided by the number of trials. Although chiefly an applied mathematician Poisson also made some significant contributions to pure mathematics, in particular to complex analysis. It was Poisson who first thought of integrating complex functions along a path in the complex plane.

Poncelet, Jean Victor. (1788-1867), French pioneer of

Projective Geometry, whose principal career was as a military engineer. He served under Napoleon in his march on Moscow, when he was left for dead; later he greatly improved the efficiency of turbines and water-wheels while Professor of Mechanics at Metz. He discovered the Principle of Duality and used Points at Infinity in order to increase the generality of geometric results such as Desargues Theorem. Poncelet was a military engineer in Napoleon's Russian campaign and was taken prisoner in 1812 during the retreat from Moscow, having been left for dead by the army. During his two years in Russia as a prisoner of war he set about reconstructing as much as he could remember of the mathematics he had learned as a student. He was able to go beyond merely reconstructing what he had been taught, to do original work. Poncelet's important contribution to mathematics was in the field of projective geometry. He was one of the first to formulate and make extensive use of the principle of duality — the equivalence of certain related geometric concepts and theorems. He was also the first mathematician to introduce imaginary points into projective geometry.

Proudman, Joseph. (b. Dec. 30, 1888; Unsworth, England; d. June 26, 1975; Verwood, England). British mathematician and oceanographer. Proudman studied mathematics and physics at Liverpool University and, after graduating in 1910, went on to do further study at Cambridge University. On completing his studies in Cambridge, Proudman returned to Loverpool (1913) in the capacity of lecturer in mathematics and remained associated with this university for the rest of his career. In 1919 he became the first professor of applied mathematics and in 1933 a chair in oceanography was created specifically for him. He remained at Liverpool, serving also as pro-vice-chancellor (1940-46), until his retirement in 1954. Proudman's abiding scientific interest was in the study of the tides and other aspects of the sea. He made fundamentally important contributions of both a practical and theoretical nature to oceanography. In 1916 Horace Lamb was asked to prepare a report on tidal research and Proudman contributed to this a paper on the harmonic analysis of tidal observations. In 1919 he founded an institute in Liverpool devoted to research on tidal phenomena, which, after several changes of name, eventually became known as the Bidston Laboratory of the Institute of Oceanographic Sciences. He worked closely on tidal phenomena with another member of the institute, Arthur Doodson. Proudman's work pioneered the mathematical study of the tides, and only as a result of his work did a unified mathematical approach to oceanography become possible. His works included *Dynamical Oceanography* (1953).

Ptolemy. (Claudius Ptolemaeus) (*fl.* 2nd century Virtually nothing is known about the life of Ptolemy. He was probably a Hellenized Egyptian working in the library at Alexandria. He produced four major works, the *Almagest*, the *Geography*, the *Tetrabiblos*, and the *Optics*. The first work — the culmination of five hundred years of Greek astronomical and cosmological thinking — was to dominate science for 13 centuries. Ptolemy naturally relied on his predecessors, especially Hipparchus, and did not particularly try to hide it. A work of such staggering intellectual power and complexity could never be created by one man alone. The basic problem he faced was to try to explain the movements of the heavens on the assumption that the universe is geocentric and all bodies revolve in perfectly circular orbits moving with uniform velocity. As the heavenly bodies move in elliptical orbits with variable velocity around a center other than the Earth some quite sophisticated geometry is called for to preserve the basic fiction. Ptolemy used three complications of the original scheme: epicycles, eccentrics, and equants. These devices worked reasonably well except that they did not lead to particularly accurate predictions. Nor did they permit Ptolemy to develop a system of the universe as a whole. He could give a reasonable account of the orbit of Mars, and of Venus, and of Mercury, and so on, taken separately, but if they were put together into one scheme then the dimensions and the periods would start to conflict. Whatever its faults the system remained intact for 1300 years until it was overthrown by Copernicus in the 15th century. In the *Geography* Ptolemy explains fully how lines of latitude and longitude can be mathematically determined. However no longitudes were astronomically determined and only a few latitudes had been so calculated. Positions of places were located on this dubious grid by reducing distances measured on land to degrees. Distances over seas were simply guessed at. As he had put the Canaries 7° east of their true position his whole grid was thrown out of alignment. The *Geography* had almost as great (and as enduring) an influence on the western world-view as the *Almagest*. Columbus might never have sailed without Ptolemy's erroneous view that Asia was closer (westward) than it really is, a view endorsed by the mapmakers contemporaneous with Columbus. The only book of Ptolemy's that is readily available today and still widely read is the *Tetrabiblos*, which is a work on astrology. The work is long and comprehensive and is probably as well argued as the case for astrology can be. It is naturalistic in that he supposes that there might be some form of physical radiation from the heavens that affects mankind. Most of the concepts and arguments of modern astrology can be traced back to this Ptolemaic work. The final major work of Ptolemy, the *Optics*, in which he sets out and demonstrates various elementary principles, is in many ways the most successful of all his works. Although he understood the principles of reflection reasonably well his understanding of refraction seems to be purely empirical. He gives tables he has worked out for the refraction of a ray of light passing from light into water for various angles. His main work was known in Greek as the *Syntaxis*; it was the Arabs who named it the *Almagest* from the Arabic definite article 'al' and their own pronunciation of the Greek work for 'great'. Such was the tribute posterity has paid to Ptolemy.

Purbach. (or Peurbach), Georg von (b. May 30, 1423; Purbach, Austria; d. Apr. 8, 1461; Vienna). Austrian astronomer and mathematician. Purbach took his name from his birthplace. He had traveled in Italy and studied under Nicholas of Cusa before becoming professor of mathematics and astronomy at the University of Vienna in about 1450. Purbach's main aim as a scholar was to produce an accurate text of Ptolemy's *Almagest.* The most common available text was that of Gerard of Cremona, which was a Latin translation of an Arabic translation and was nearly 300 years old. Purbach began by writing a general introduction to Ptolemy that described accurately and briefly the constructions of the *Almagest.* Unfortunately he died before he could embark on the translation. His place was taken by his pupil,

Regiomontanus, who completed a textbook begun by Purbach but failed to produce the edition and translation of Ptolemy so much wanted by Purbach. One of his most significant works, the fruit of much observational and theoretical work, was a very thorough table of lunar eclipses, which he published in 1459. Purbach wrote a textbook, *Theorica novae planetarum,* which became an influential exposition of the Ptolemaic theory of the solar system, a theory whose influence lasted until Tycho Brahe finally disproved the existence of the solid spheres postulated by Ptolemy. Such was the accuracy of Purbach's set tables that they were still in use almost two hundred years later. He also compiled a table of sines, using Arabic numerals, and was one of the first to popularize their use instead of chords in trigonometry.

Pythagoras (b. *c.* 580 BC; Samos, Greece; d. *c.* 500 BC; Metapontum, now in Italy). Greek mathematician and philosopher. All that is known of the life of Pythagoras with any certainty is that he left Samos in about 520 BC. to settle in Croton (now Crotone) in southern Italy and as a result of political trouble made a final move to Metapontum in about 500. In Croton Pythagoras established his academy and became a cult leader. His community was governed by a large number of rules, some dietary, such as those commanding abstinence from meat and from beans, and others of obscure origin, such as the commands not to let a swallow nest under the roof or not to sit on a quart measure.The movement was united by the belief that "all is number." While the exact meaning of this may be none too clear, that it led to one of the great periods of mathematics is beyond doubt. Not only were the properties of numbers explored in a totally new way and important theorems discovered, of which the familiar theorem of Pythagoras is the best example, but there also emerged what is arguably the first really deep mathematical truth — the discovery of irrational numbers with the realization of the incommensurability of the square root of 2.

Q

Quetelet. (Lambert) Adolphe (Jacques) (b. Feb. 22, 1796; Ghent, now in Belgium; d. Feb. 17, 1874; Brussels). Flemish astronomer, mathematician, and sociologist. Quetelet, the son of a municipal official, was educated at the Lycee de Ghent and, when only 19, was appointed as an instructor in mathematics at the Royal College in Ghent. In 1819 he moved to Brussels to take the chair of mathematics at the Athenaeum, a post he held until his appointment in 1828 as director of the newly established Royal Observatory. It is, however, as a statistician and not as an astronomer that Quetelet is remembered. In this field he appeared to have an obsession with the collection and analysis of the variations found in natural phenomena. Thus, in 1825, he began by preparing a table of births and deaths in Brussels that was later extended to cover the whole of Belgium. He soon branched out to cover the statistics of crime and tried to show their relationship to such variables as sex, age, climate, and education. In various influential works Quetelet argued for the use of statistics in the establishment of a social science and for the discovery of social laws. This idea, together with his concept of the 'average man' (*l'homme moyen*) caused much controversy.

R

Rabi, Isidor Isaac (b. July 29, 1898; Raymanov, now in the Soviet Union). Austrian-American physicist. Rabi's parents emigrated to America while he was still young and he grew up in a Yiddish-speaking community in New York where his father ran a grocery store. He was educated at Cornell, graduating in 1919, and Columbia, where he obtained his PhD in 1927. After two years in Europe he returned to Columbia where he spent his whole career until his retirement in 1967, being appointed professor of physics in 1937 and the first University Professor (a position with no departmental duties) there in 1964. While in Germany (1927) Rabi had worked under Otto Stern and was impressed with the experiment Stern had performed with Walter Gerlach in which the use of molecular beams led to the discovery of space quantization (1922). Consequently Rabi began a research program at Columbia where he invented the atomic— and molecular-beam magnetic-resonance method of observing atomic spectra, a precise means of determining the magnetic moments of fundamental particles. Using his techniques after World War II, experimentalists were able to measure the magnetic moment of the electron to nine significant figures, thus providing a powerful tool for the testing of theories in quantum electrodymamics. The method had wide applications to the atomic clock, to nuclear magnetic resonance, and to the maser and laser. For this work Rabi was awarded the Nobel Prize for physics in 1944. During the war Rabi worked on the development of microwave radar. In the postwar years he was a member of the General Advisory Committee of the Atomic Energy Commission, serving as its chairman (1952-56) following the resignation of J. Robert Oppenheimer. As a member of the American delegation to UNESCO he originated the movement that led to the foundation of the international laboratory for high-energy physics in Geneva known as CERN.

Ramanujan, Srinivasa. (1887-1920). Indian mathematician who was largely self-taught; he contributed greatly to the theory of numbers and the theory of functions using mainly intuitive methods, and is famed for his skills in manipulation of series. He lost a scholarship to Madras University by concentrating on mathematics to the exclusion of other subjects, and for some time existed on private charity before obtaining a clerical post. After corresponding with G.H. Hardy, he obtained a research scholarship at Cambridge in 1914, and in 1919 was the first Indian to be elected a Fellow of the Royal Society, before returning to India much weakened by a mysterious disease then thought to be tuberculosis. Ramanujan is often regarded as one of the romantic figures of mathematical history, and his work is only now being fully digested. Ramanujan was born in an obscure Indian village and worked as a junior clerk in Madras. He had only an elementary mathematical education at school and yet was able to rediscover much of the mathematics that had originally been discovered by some of the best mathematicians of the 19th century and to arrive at many completely new results. After a two-year fellowship at the University of Madras he was enabled to go to England to study with G. H. Hardy in Cambridge through a scholarship from the Madras local government and a grant from Cambridge University. Hardy has recorded his astonishment on receiving Ramanujan's first letter and reading through the hundred or so theorems it contained. Ramanujan's work was entirely in the fields of real analysis and number theory. He had no interest in applied mathematics. He had astonishing facility at mental calculation and arrived at many of his results by working out large numbers of examples. While much of Ramanujan's work was great originality and depth there were large gaps in his knowledge; for example, he knew almost nothing of complex analysis. The work that was the greatest stimulation to later mathematicians was on the partition of numbers into summands.Ramanujan was a devout Hindu and strict vegetarian. His religious scruples almost prevented him going to England. He returned to Indian in 1919 and died there of tuberculosis the next year.

Ramsey, Frank Plumpton (b. Cambridge U K, 22.2.1903; d. Cambridge, 19.1.1930). British philosopher, mathematician and economist. Having read mathematics at Trinity College, Cambridge, he became a fellow of King's College. Ramsey was a genius who made outstanding contributions to three quite different disciplines: philosophy, mathematics and economics. In each case the arguments or field of investigation that he opened up came to be extensively developed only some 30 years after his death at the tragically early age of 26. In philosophy, Ramsey's work was concerned largely with the nature of various kinds of knowledge. He wrote a number of important papers, several of them stimulated by Bertrand Russell's and Whitehead's *Principia Mathematica* (1913) or by Wittgenstein's *Tractatus* (1921). In mathematics, he established a pair of related theorems about infinite sets, which gave rise to an extensive new field in mathematics subsequently called 'Ramsey theory'. In economics, Ramsey's two papers each pioneered a branch of theory: optimal taxation or (more broadly) the theory of 'second best' and optimal savings or accumulation. 'A contribution to the theory of taxation' (1927) considered the question of what tax rates should be applied to different commodities in the raising of a given amount of revenue, if the tax-payers' loss of well-being is to be minimized — the answer being that rates should not be uniform but should vary inversely with the elasticity of demand for a commodity. 'A mathematical theory of saving' (1928) used the calculus of variations to derive a theorem about the proportion of income that a community should save. The proportion is shown, plausibly enough, to vary positively with the scope for increasing society's well-being by higher consumption in the future, and inversely with the value attached to an increment of consumption in the present. J. M. Keynes, 'F. P. Ramsey' in *Essays in Biography* (London, 1933); D.H. Mallor (ed.), *F. P. Ramsey: Foundations — Essays in Philosophy, Logic, Mathematics and Economics* (London, 1978).

Regiomontanus (Johann Müller). (b. June 6, 1436; Königsberg, now Kaliningrad in the Soviet Union; d. July 6, 1479; Rome). German astronomer and mathematician. Regiomontanus, as befitted a Renaissance humanist, changed his name, Johann Müller, for a Latin version of the name of his home town Königsberg. His father was a miller. He was educated at Leipzig and

Vienna where he was a pupil of Georg von Purbach. One of the ambitions of Purbach had been to produce a good text of Ptolemy's *Almagest* based on the original Greek rather than translations from Arabic at third or fourth hand. He had intended to go to Italy in quest of manuscripts of Ptolemy and other ancient scientists with the great Greek scholar Cardinal Bessarion. The cardinal's death in 1461 allowed Regiomontanus to take his place and spend six years in Italy searching for, translating, and editing manuscripts. After his return he settled in Nuremberg where his wealthy benefactor, Bernard Walther, built him an observatory and provided him with instruments. In 1475 he was called to Rome by the pope, Sixtus IV, to help in the reform of the calendar but died of the plague (or, possibly, poison) in 1476. Regiomontanus was one of the key figures of 15th century science. In the 1460s he wrote *De triangulis* (On Triangles), a work not printed until 1533 but that was, together with *Tabulae directionum* (1475; Tables of Direction), the main channel for the introduction of modern trigonometry into Europe. In the latter work he broke away from the ancient tradition of chords and instead gave tables of sines for every minute and tangents for every degree.

Reinhold, Erasmus. (b. Oct. 22, 1511; Saalfeld, now in East Germany; d. Feb. 19, 1553; Saalfeld). German astronomer and mathematician Reinhold was one of the outstanding mathematical astronomers of the 16th century, his only rival being Copernicus. Nothing is known of his early life. He was a student at the University of Wittenberg, where he subsequently became both dean and professor of mathematics. He published two widely influential commentaries on astronomical works by Georg von

Purbach and by Ptolemy. He knew Copernicus personally and had an extremely high opinion of his work, although he does not seem to have viewed Copernicus's heliocentric theory of the solar system as anything more than an interesting hypothesis. Reinhold computed his astronomical tables using the methods that Copernicus had incorporated into his *De revolutionibus.* The resulting *Tabulae prutenicae* (1551; Prutenic Tables, as they were called) became widely used.

René Descartes. René Descartes was born in 1596 in a village near Tours called La Haye, which is now known as Haye-Descartes to honor its most Famous son. He was sickly as a child. Like so many people who are brilliant but in poor health, Descartes devoted most of his time to study. In the general framework of French culture he is known especially for his *Discours sur la méthode (Discourse on Method),* in which he tried to build a complete philosophical system starting from scratch, with the single assumption *Cogito, ergo sum:* I think, therefore I am (or in the vernacular, *Je pense, donc je suis*). The book was published in 1638, four years before the birth of Isaac Newton, and became an instant success, establishing Descartes as one of the giants of French culture. He dabbled in cosmology and came up with a concept of the Universe as being composed of rotating matter, which he called vortices; in the case of the earth, he laid it to rest, in the center of one of his vortices, which performed the movement carrying the earth about the sun. It was a clever way of putting the earth into orbit about the sun without actually embracing the Copernican heresy (and the Jesuits, with whom he had studied as a young man, are traditionally famous for too-clever schemes of this type). Unfortunately for

Descartes, within a generation his cosmology would be abolished forever by Newton. Descartes's greatest and most durable contributions to science were in mathematics. In algebra he imposed the system of using letters from the beginning of the alphabet to represent constants and those at the end for variables. He invented Cartesian (after the Latin version of his name, Cartesius) coordinates and extended his research to the formulation of analytic geometry, which is still an imperishable monument to his genius. Watching the tip of a branch outside his window move across the panes. He realized that he could specify the location of the branch in the window by noting how many panes away from the lower left corner pane the branch appeared and how many panes it appeared to the right. If the pane occupied by the branch was two panes over and three panes up, its location could be abbreviated as (2,3). Descartes assumed that the location of any point on a plane could be specified by a pair of numbers relative to a fixed pair of crossed lines. If we think of one line as horizontal and the other line as vertical, we can designate the two numbers x and y. In addition, we can designate one of the horizontal and one of the vertical lines as the x axis and the y axis, respectively. The number x can be measured along the x axis, and gives the distance from the y axis; the number y can be measured along the y axis, and gives the distance from the x axis. If we were dealing with a point in a three-dimensional volume, we would require three numbers or coordinates to locate the point. This system of coordinates transformed mathematics because it showed that anything said about a geometrical point could be translated into a numerical statement. Descartes's grand vision of a geometrical Universe led him to disdain efforts to solve particular problems of celestial mechanics. He was contemptuous of Galileo's work, particularly his law of freely falling bodies and his experiments with the pendulum. Galileo's conclusions depended on the existence of a vacuum, which could not exist in Descartes's Universe, where all spaces were actually filled with minute particles. Indeed, Descartes's major contribution to mechanics may have been the philosophical stimulus he provided to other physicists such as Christian Huygens. Descartes believed light to be a force transmitted at infinite speed by the particles in between visible objects in the Universe. Descartes's model "for light itself was the blind man's cane, which instantaneously transmits impulses from the objects it meets and enables the man to `see.'"* Descartes also discovered a way to describe mathematically how sunlight is refracted and reflected in raindrops to form a rainbow. However, he was less successful in explaining what causes the rainbow's colors. Our selection is from a work which, although written in French, bore a Latin title, *Principia philosophiae (Principles of Philosophy).*

Reynolds, Osborne (b. Aug. 23, 1842; Belfast; d. Feb. 21, 1912; Watchet, England). Irish engineer and mathematician. After gaining experience in workshop engineering, Reynolds went to Cambridge University, graduating in 1867. In 1868 he became the first professor of engineering at Owens College, Manchester. Reynolds made valuable contributions to hydrodynamics and hydraulics. He is best-known for the *Reynolds number,* which he introduced (1883-84) as a dimensionless parameter that determines whether fluid flow is smooth or turbulent. His studies of condensation and heat transfer led to a radical revision of boiler and condenser design. He formulated theories of lubrication and of turbulence, helped to develop the idea of group velocity of waves, explained how radiometers work, and determined the mechanical equivalent of heat. Reynolds became a fellow of the Royal Society in 1877 and a Royal Society medalist in 1888.

Rhäticus, or Rheticus (b. Feb. 16, 1514; Feldkirch, Austria; d. Dec. 5, 1576; Kassa, Hungary). Austrian astronomer and mathematician. Rhäticus was born Georg Joachim von Lauchen but he adopted his name from the Austrian district, Rhaetia, in which he was born. In 1537 he was appointed professor of mathematics and astronomy at the University of Wittenberg in Germany. Rhäticus played a considerable part in the popularization of Copernicus's heliocentric view of the solar system. He traveled to Poland in 1539 specifically to study with Copernicus and actually published the Copernican theory before Copernicus himself in his *Narratio prima de libris revolutionum Copernici* (1541 The First Account of the Revolutionary Book by Copernicus). He persuaded Copernicus to complete his work for publication and this appeared as *De revolutionibus orbium coelestium* (1543). In addition Rhäticus wrote a major treatise *Opus palatinum de triangulis* in which he gave tables of trigonometric functions accurate to ten decimal places. This treatise was not, however, published until after his death.

Ricci, Matteo. (b. Oct. 6, 1552; Macerata, Italy; d. May 11, 1610; Peking). Italian astronomer and mathematician. Ricci, a Jesuit, was the son of a pharmacist. He received a rigorous education from the Jesuits in Rome, including classes in mathematics and astronomy from Christopher Clavius. After a few years' missionary work in India he arrived in Macao in 1582 to wait his chance to gain admission to the centralized xenophobic Ming China. The policy of the Celestial Empire incorporated a generalized contempt for all foreigners as uncultured barbarians together with the not unreasonable claim that the least offensive place for such creatures was their own country. If foreigners themselves were despised, foreigners who presumed to possess a superior religion and culture were beyond comprehension. Thus Ricci's options were very limited, forcing him to gain entry in the only practical way, that is, by becoming an unofficial Chinese. To gain respect and influence he had to go further and become a Chinese scholar. It took 20 years for Ricci to reach Peking, for first he had to master Chinese culture, language, and literature. Once accepted as a scholar he found many in the Mandarinate who were keen to learn western mathematics and astronomy. Thus he translated the first six books of Euclid into Chinese in 1607, together with works of his teacher Clavius. The Chinese scholars appreciated the many fields in which Ricci was their superior. This had the effect that Chinese science after 1600 ceased to exist as an independent tradition, for although the door could be closed against the Bible and Christianity, it was impossible to shut one's mind to Euclid. Thus Chinese science, because of Ricci, at last became part of universal science. One ironical aspect of this was that Ricci introduced the astronomy of Ptolemy into China with all its medieval accretions just as it was being rejected in Europe. Ricci had one further advantage that gained for him his ten years' residence in Peking. He had brought with him a magnificent clock with 'self-sounding bells' for the emperor Wan Li. As the Chinese scholars were largely ignorant of their own horological tradition, Ricci and the Jesuits were charged with its installation, its upkeep, and the training of eunuchs to care for it. Such was the respect the emperor felt for Ricci, though they never met, that on Ricci's death, land was made available on which his tomb could be built.

Richard of Wallingford. (b. 1291/92; Wallingford, England; d. May 23, 1336; St. Albans, England). English astronomer and mathematician. Richard's father was a blacksmith and after his death Richard was adopted by the prior of Wallingford. He was at Oxford University as a student from 1308 to 1314 and taught there from 1317 to 1326 before becoming the abbot of St. Albans. He is thought to have contracted leprosy in early life and there is a manuscript illustration of him in the British Museum that show him with a spotty or scarred face. Oxford at this time had gone through a minor renaissance. There were a number of scholars including Richard who were profoundly aware of the limitations imposed by traditional mathematical methods in dealing with virtually any problem of physics. It was Richard who introduced trigonometry into England in its modern form and in a series of manuscripts he produced the basic texts that could have initiated a mathematical revolution. (He was, however, two centuries too soon. The political troubles of the next 200 years and the black death were sufficient to smother any premature intellectual birth.) He was not just a theoretical mathematician for he designed and made his own instruments and, above all, he designed a marvelous clock for his abbey. It has been suggested that he introduced the work 'clock' into the English language, from the Latin 'clocca' for bell. His clock, the plans for which survive, probably predated that of Giovanni de Dondi in the use of an escapement.

It showed the position of the Sun, the Moon, the stars, the state of the tide — in fact, it seemed, like most of the medieval clocks, to do just about everything except tell the time.

Richter, Burton (b. Mar. 22, 1931; New York City). American physicist. Richter, joint winner of the 1976 Nobel Prize for physics, studied first at the Massachusetts Institute of Technology, gaining his BS in 1952 and his PhD in physics in 1956. His interest in the physics of elementary particles took him subsequently to Stanford University's high-energy physics laboratory where he became a member of the group building the first pair of electron-storage rings. In this machine, intense beams of particles were made to collide with each other in order to study the validity of quantum electrodynamic theory. In the 1960s, Richter designed the Stanford Positron Electron Accelerating Ring (SPEAR), which was capable of engineering collisions of much more energetic particles. It was on this machine that, in November 1974, Richter and his collaborators created and detected a new kind of heavy elementary particle, which they labeled psi (Ψ). The discovery was announced in a 35-author paper (typical of today's high-energy-research teams) in the journal *Physical Review Letters*. The particle is a hadron with a lifetime about one thousand times greater than could be expected from its observed mass. Its discovery was important as its properties are consistent with the idea that it is formed from a fourth type of quark, thus supporting Sheldon Glashow's concept of 'charm'. Almost simultaneously, another group led by Samuel Ting 2000 miles away at the Brookhaven Laboratory, Long Island, made the same discovery independently in a very different experiment. Richter and Ting met to discuss their findings, and confirmation came quickly from other laboratories when they knew the energy of the new particle and were able to tune their own machines accordingly. Ting called the new particle J; it is now usually referred to as the J/psi in recognition of the simultaneity of its discovery. The discovery led to the finding of many other similar particles as a 'family' and has stimulated new attempts to rationalize the underlying structure of matter. Within only two years, Richter and Ting were to be the recipients of the Nobel Prize for physics. Richter has been a full professor at Stanford University since 1967, taking a sabbatical year at the European Organization for Nuclear Research (CERN) in Geneva (1975-76).

Riemann, (Georg Friedrich) Bernhard. (b. Sept. 17, 1826; Breselenz, now in West Germany; d. July 20, 1866; Selasca, Italy). German mathematician. Before studying mathematics in earnest, Riemann studied theology in preparation for the priesthood at his father's request. Fortunately he was able to persuade his father, a Lutheran pastor, that his real talents lay eslewhere than in theology. He attended the University of Göttingen and his mathematical abilities were such that his doctoral thesis won the rarely given praise of Karl Friedrich Gauss. After gaining his doctorate Riemann worked on the inaugural lecture necessary in order to gain the post of *Privatdozent* at Göttingen and this too gained Gauss's praise. Eventually Riemann succeeded his friend, Lejeune Dirichlet, as professor of mathematics at Göttingen in 1859 but by then his health had begun to decline and he died of tuberculosis while on holiday in Italy. Riemann's work ranges from pure mathematics to mathematical physics and he made influential contributions to both. His work in analysis was profoundly important. The

Riemann integral is a definite integral formally defined in terms of the limit of a summation of elements as the number of elements tends to infinity and their size becomes infinitesimally small. One of Riemann's most famous pieces of work was in geometry. This was initiated in his inaugural lecture of 1854 that so impressed Gauss, entitled "Concerning the Hypotheses that Underlie Geometry." What Riemann did was to consider the whole question of what a geometry was from a much more general perspective than anyone had previously done. Riemann asked questions, such as how could concepts like curvature and distance be defined, in such a way as to be applicable to geometries that were not Euclidean. Janos Bolyai and Nikolai Lobachevsky (and, at the time unknown to everyone, Gauss) had developed particular non-Euclidean geometries, but Riemann went further and opened up the possibility of a range of geometries different from Euclid's. This work had far-reaching consequences, not just in pure mathematics but also in the theory of relativity. Riemann was also interested in applied mathematics and physics and was a co-worker of Wilhelm Weber.

Robert Hooke. Robert Hooke was in many ways one of the most pitiable men of talent in the history of physics. He was nasty-tempered, vengeful and quarrelsome as well as not above trying to pass off the discoveries of others as his own. Hooke was born on 18 July 1635, on the Isle of Wight, off the northern shore of what the British call the English Channel. Hooke's father was a clergyman. As a boy, Hooke was plagued by sickness and scarred by smallpox to the point of ugliness. Yet he showed an extraordinary mechanical ability very early in life. By the time he was eighteen he had been accepted at

Oxford, where he supported himself by waiting on tables, a traumatic humiliation that seems to have had a permanent effect on his personality. He became a coworker of Robert Boyle, whose celebrated air pump owed much to Hooke's perfectionist skill as a tinkerer. As a scientist Hooke made useful but not conclusive contributions to the wave theory of light and the theory of gravity. His interests ranged from these matters to pre-Daltonian atomic studies, astronomy, earthquakes and the physics of spring mechanisms. In biology he was an expert microscopist, and his book *Micrographia*, published in English, its Latin title notwithstanding, was illustrated with remarkable pen-and-ink drawings of microscopic objects. His microstudies of the composition of cork led him serendipitously to suggest the use of the word cell (meaning a tiny, bare room, like a monk's cell), and the word survived as the name for living cells. Hooke devoted much of his early career to the study of solid springs. In his lectures, he demonstrated that a doubling of the weight at the end of a copper spring would double the stretching of the spring itself. He soon turned his attention to the "spring" or resistance of the air itself and discovered that it is not constant but increases as the air is compressed. He found that increased pressure "stiffened" the air. More precisely, it follows from Boyle's law that the spring of the air varies inversely with its volume at a constant temperature. Hooke proposed that the spring of the air results entirely from the collisions of atoms with the surface of their container. Yet he was reluctant to cut the atoms loose from the fixed positions decreed by Newton, so he permitted his atoms only to jiggle to and fro, never allowing them to venture far enough to encounter each other. For one air atom to transmit its force to its neighbor, Hooke relied on Newton's ether, but declared that it too was made up of atoms which bounce back and forth between the air atoms. He pointed out that when the volume of a gas is reduced at a constant temperature, the speeds of the atoms do not change but they strike the walls of the container more frequently having less distance to go. Hooke then considered what type of collisions between atoms could hold an object such as a copper spring together. Not surprisingly, he assumed that a body is "pushed together" by the impacts of ether particles on its surface. Hooke then tried unsuccessfully to bring gravity into the fold of his grand scheme reasoning that two bodies immersed in a sea of invisible ether particles running wildly in all directions will be pushed together because they shelter each other from impacts on their near faces but not their distant faces. Unfortunately for Hooke, this reasoning was flawed; the random collisions between the ether atoms would make it impossible for the near faces to be protected and the two bodies pushed together. Besides his substantial but not overpowering skills as an independent scientist, Hooke was a tireless operator, in the modern, manipulative sense of the word. From 1677 to 1683 he was a powerful secretary of the Royal Society, and from 1662 until his death forty-one years later he was the Society's "curator of experiments." His was the only paid job in the Society, and it gave Hooke a kind of absolute administrative power over his colleagues—including his betters, such as Isaac Newton.

Roger G. Boscovich. Roger Boscovich was an eighteenth-century Jesuit mathematician who pioneered the development of atomic science and was among the earliest Italians to adopt Newton's theory of gravitation. Born in 1711 at Ragusa in Dalmatia, he entered the

Society of Jesus at the age of fifteen and studied mathematics and physics at the Collegium Romanum. After being appointed professor to his college in 1740, he published dissertations on a variety of subjects including the orbit of Mercury, the Aurora Borealis, terrestrial gravitation, the mathematics of the telescope, comets, tides, and trigonometry. His famous work in atomic theory, *Theoria philosophiae naturalis redacta ad unicam legem virium in natural existentium,* was published in 1771. A diplomatic visit to England resulted in his election to the Royal Society. The 1773 suppression of his order convinced him to accept the invitation of the French king to become the director of optics for the navy where he remained for a decade until returning to Italy for good to live in Bossano. Boscovich hypothesized a set of so-called *point centers* that allowed physicists to abandon their old theory of a variety of different solid atoms. The basic components of matter, in his view, were identical, and matter was determined by certain relationships of space about these point centers. His writings, ranging from the astronomical and mathematical to the world of atoms, was one of the earliest harbingers of the unification of the infinite and infinitesimal in the structure of the Universe. To account for the different separations among atoms in a gas, liquid, and solid, he modified the force law to include equally many points of equilibrium, where the force changes from repulsion to attraction, with a brief stasis at zero. In his scheme, the distance between two atoms was plotted along the horizontal axis, the force on the vertical axis, with attractions indicated below the horizontal axis and repulsions above the horizontal. The curve crossed the coordinate at a point of equilibrium, of zero force. Boscovich seemed to recognize two kinds of equilibria, stable and unstable. Segments where the force curve ascends from left to right as it passes through the equilibrium point are unstable, since if an atom were put at such a point the slightest inward displacement would position it in such a way that the force would draw it further in, and the least outward change would place it where the force would push it farther out. Such equilibrium is like that of a free-standing pole (such as a broomstick): it cannot long endure in a continually perturbing environment. Only at a point of equilibrium where the force is repulsive at slightly smaller distances and attractive at slightly greater distances (so that the curve of force descends from left to right) is there a physically possible equilibrium, like that of a pole lying flat. Clearly, the closer the Newtonian theory of matter approximated experiment, the more complex it was seen to be.

Roche, Edouard Albert. (b. Oct. 17, 1820; Montpellier, France; d. Apr. 18, 1883; Montpellier). French mathematician. Roche studied at the University of Montpellier where he obtained his doctorate in 1844. After further study in Paris he returned to Montpellier in 1849 and served as professor of pure mathematics from 1852 until his retirement in 1881. Roche's name is still remembered by astronomers for his proposal in 1850 of the limiting distance since named for him. He calculated that if a satellite and the planet it orbited were of equal density then the satellite could not lie within 2.44 radii, the *Roche limit,* of the larger body without breaking up under the effect of gravity. As the radius of Saturn's outermost ring is 2.3 times that of Saturn it was naturally felt that the rings could well consist of broken-down fragments of a former satellite that had transgressed the forbidden limit. It is now thought, however, that the Roche limit has prevented the fragments from aggregating into a satellite. Roche later worked on the nebular hypothesis of Pierre Simon de Laplace, submitting it to a rigorous mathematical analysis and concluding in 1873 that a rapidly rotating lens-shaped body was in fact unstable. He also published work on the structure and density of the Earth and produced a generalization of Taylor's theorem, much used in mathematics.

Roger Penrose. Roger Penrose, a British mathematician and theoretical physicist, was born in 1931 in Colchester, England. His interest in science was doubtless fueled by his father, the geneticist Lionel Penrose. Roger's mathematical brilliance showed early both as an undergraduate at University College, London, and as a doctoral candidate at Cambridge where he earned his Ph.D. in 1957. After working as a lecturer and researcher at London and Cambridge (and also at Princeton, Syracuse, and Texas), Penrose settled down permanently in 1973 as professor of mathematics at Oxford. Penrose is best known for his work in general relativity theory especially as it applies to the gravitational collapse of stars. He has used his formidable mathematical talents to explore theoretically the properties of black holes. Penrose and Stephen Hawking proposed a theorem which holds that any object falling inside the event horizon of a black hole will be cut off from the outside world because it is then surrounded by a region of curved space-time that is so strongly curved that not even light can escape. Alternatively, it is the surface of the black hole where objects spiral inward at the speed of light. A light ray traveling outward at the event horizon remains static since its velocity is exactly equal to the speed of escape. Inside the event horizon, objects flow toward the center of the black hole faster than the speed of light. As a result, all communication with the outside Universe is cut off. Penrose showed that all black holes ultimately collapse to singularities-centers of infinite density and pressure and zero volume where the laws of physics break down. That we cannot calculate the forces exerted by the singularity has prompted many physicists to question whether the general theory of relativity is not somehow flawed since it becomes impotent near the singularity. This breakdown of classical general relativity occurs because the singularity is assumed to be a point that contains all the mass of the black hole. But Stephen Hawking's work showing that black holes can evaporate may provide a way for avoiding the problems of the point singularity since all energy falling down into a black hole will eventually be thermally reradiated into the universe. Even though the time needed for a ten-solar-mass black hole to evaporate is abcut 10^{66} years, the singularity would no longer be a point of infinite density and zero volume. It would instead have a finite, albeit extremely small size. In effect, we would substitute the probabilities of quantum theory for the uncalculable densities and volumes predicted by Einstein's general theory of relativity. Like Hawking, Penrose has studied the quantum mechanics of spining black holes and demonstrated that an object falling into the ergosphere-the region of swirling space between the outer static surface and the event horizon-can divide into two parts, with one part continuing its inevitable journey toward the singularity and the other escaping from the ergosphere. The latter piece would take with it some of the black hole's rotational energy, emerging with more energy or mass than the original falling object. It would thus be possible to harness the energy of the black hole.

Rudolf Clausius. Rudolf Julius Emmanuel Clausius was born in 1822 in a Pomeranian city that is now part of Poland. He was the son of a preacher and, like Carnot, received his early education at home from his father. Clausius studied physics and mathematics at the University of Berlin and received his doctorate at Halle in 1847. He was appointed professor of physics at the University of Zurich in 1855 due in part to his 1850 paper which set out his theory of heat and established the basis for the modern study of thermodynamics. Clausius died in Bonn in 1888. He was a mathematician and theoretical physicist who, together with Janes Clerk Maxwell and Ludwig Boltzmann, contributed much to the development of the kinetic theory of gases. He was the first to suggest that when an electric current is passed through a solution, the molecules are dissociated by the current. This concept, revolutionary for its day, was not generally accepted until presented again by Svante Arrhenius a generation later. In 1860, reconsidering the earlier work of Carnot, Clausius discovered the principle which he named entropy and which he defined as the ratio between the heat content of a closed system and its absolute temperature, a ratio that invariably increase and never decreases. His formulation of the idea, the second law of thermodynamics, was so clear that he is often credited with its discovery, although Maxwell, Boltzmann, and Kelvin share that honor. Clausius realized that the entropy of an entire system is the sum of the entropies of its parts. He also found that the entropy of a closed system undergoing a process increases if and only if the process is irreversible. From these two conclusions it followed that two systems would be in thermal equilibrium if and only if they have the same values for the rate of change of temperature. This temperature, defined universally without dependence on a particular thermometer, is called the absolute temperature T; it is defined by the formula $T = dQ/dS$ *or* $dS - DQ/T$ mentioned by Clausius where dS is the infinitesimal change in the system's entropy S and dQ is the infinitesimal heat gained or lost. In this selection Clausius reformulated the second law of thermodynamics as the principle that the total entropy of a closed system can never decrease. Since the only perfectly closed system is the entire Universe, the concept of the eventual "heat death" of the Universe follows from it, a condition in which a constantly rising entropy leads to complete temperature equilibrium. As od now it is not clear, however, whether the laws of thermodynamics are applicable throughout the Universe. Clausius' paper on the second law of thermodynamics was originally published in the *Annalen der Physik und Chemie* in 1850.

Russell, Henry Norris (b. Oct. 25, 1877; Oyster Bay, New York; d. Feb. 18, 1957; Princeton, New Jersey). Russell, the son of a Presbyterian minister, was a brilliant scholar at Princeton, graduating in 1897 and obtaining his PhD in 1899. He spent the period 1902-05 as a research student and assistant at Cambridge University, English, returning then to Princeton where he served as professor of astronomy from 1911 t 1927 and director of the university observatory from 1012 to 1947. He was also a research associate at the Mount Wilson Observatory in California (1922-42) and, after his retirement, at the

Harvard and Lick observatories. Russell's great achievement was his publication in 1913 of a major piece of research contained in what is now called the *Hertzsprung-Russell diagram* (h-R diagram). The same results had in fact been published earlier and independently by Ejnar Hertzsprung with little impact. Russell's work was based upon determinations of absolute magnitudes, i.e. intrinsic brightness, of stars by the measurement of stellar parallax. His measurement technique was developed in collaboration with Arthur Hinks while he was at Cambridge and involved photographic plates, then a fairly recent scientific tool. He found that values of absolute magnitude correlated with the spectral types of the stars. Spectral type was derived from the Harvard system of spectral classification as revised by Annie Cannon and indicated surface temperature. A graph of absolute magnitude versus spectral type produced the H-R diagram and showed that the majority of stars lie on a diagonal band, now called the 'main sequence', in which magnitude increases with increasing surface temperature. A separate group of very bright stars lie above the main sequence. This meant that there could be stars of the same spectral type differing enormously in magnitude. To describe such a difference the now familiar terminology of 'giant' and 'dward' stars was introduced into the literature. The most obvious feature of the diagram for Russell, however, was that it was not completely occupied by stars. This led him to propose a path of stellar evolution, which he put forward in 1913 at the same time as the diagram. He argued that stars evolve from hot giants, pass down the main sequence and end as cold dwrfs. The mechanism driving the change was that of contraction. The bulky giants of spectral type M contract and with the resulting rise of temperature move leftward in the diagram, gradually becoming B0type dwarfs. But at some stage the contraction and density become too great for the gas laws to apply and the star cools, slipping down the main sequence and evolving finally to an M-type dward. By 1926, however, Ethur Eddington could talk confidently of the overthrow of the 'giant and dward theory'; it was too simple to fit the growing data on the distribution of mass and luminosity among the different spectral types of stars. Although Russell's evolutionary theory quickly fell from favor the H-R diagram has continued to be of enormous importance and the start for any new theory of stellar evolution. Eclipsing binary stars, such as Algol, the 'winking demon', were also of great interest to Russell. He devised methods by which both orbital and stellar size could be determined, and which became widely used. He also analyzed the variations in light output of a large number of eclipsing binaries, which again became invaluable to later researchers. Another major line of research for Russell was his investigation of the solar spectrum, which began as a result of the publication in 1921 of the ionization equation of Meghnad Saha (q.v.). The Saha equation was tested and modified by Russell, using the solar spectrum, and was then used by him to calculate the abundance of the chemical elements in the Sun's atmosphere. He realized that the abundances in other stars could also be calculated from their spectra. He showed that the abundance of elements within the Sun itself could be found and in 1929 published the first reliable determination of this, demonstrating surprisingly that 60% of the Sun's volume was hydrogen. Although this was an underestimate, as Donald Menzel was later able to show that a figure of over 80% was more accurate, it dide pose the problem as to why the Sun, and presumably other stars too, should contain so much hydrogen. The answer to this question was given in the version of the big-bang theory proposed by George Gamow.

Russell, Lord Bertrand Arthur William (1872-1970), English philosopher, logician, and mathematician, noted for his ground-breaking work in mathematical logic and the foundations of mathematics. He discovered RUSSELL'S PARADOX in the axiomatization of set theory proposed by Frege, and communicated it to the latter just as the second volume of his major work was about to go to press. He obtained a lectreship at Cambridge in 1910, but was dismissed from the post, and later jailed, for making pacifist statements during the First World War. Afterwards, he taught at Harvard, the National University of Peking, the University of Chicago, and UCLA, and won many prizes, including the Nobel Prize for Literature. Russell was orphaned at an early age and brought up in the home of his grandfather, the politician Lord John Russell. He was educated privately before attending Cambridge University (1890), from which he graduated (1893) in mathematics. In 1895 he became a fellow and lecturer at Cambridge. His work after1920 was mainly devoted to the development of his philosophical and political opinions. He became well known for his popularization of many areas of philosophy and also, in worked such as *The ABC of Atoms* (1923) and *The ABC of Relativity* (1925), of the new trends in scientific thought. For his writings he was awarded the 1950 Nobel Prize for literature. He succeeded his brother to become the 3rd Earl Russell in 1931. Throughout much of his life Russell was an intense advocate of pacifism and during World War I he was imprisoned for expressing these views. Later, in the 1950s and 60s, he became a central figure in the movements criticizing the use of the atomic bomb, leading demonstrations and mass sit-downs and becoming president of the British Campaign for Nuclear Disarmament in 1958. At Cambridge Russell became interested in the relatively new discipline of mathematical logic in which he was to be a pioneer. With Giuseppe Peano he was one of the few to recognize the genius of Gottlob Frege and his new system of logic. In 1902 he wrote to Frege, presenting what is now known as *Russell's paradox,* and asking how Frege's system would deal with it. (Unfortunately, as Frege acknowledged, the system could not accommodate it.) The paradox is one of the paradoxes of set theory and rests on the (then ill-defined) notion of a set. Some sets are members of themselves (the set of all sets is an example because it is itself a set; the set of cats is not an example, as it is not itself a cat). Consider the set of all sets that are not members of themselves: is it a member of itself? If it is, it is not and vice versa. To avoid such paradoxes Russell formulated his logical theory of types. In 1903 he began his collaboration with A.N. Whitehead on their ambitious, if not entirely successful, project of placing mathematics on a sound axiomatic footing by deriving it from logic. This culminated in the publication of *Principia Mathematical* (1910, 1912, 1913), containing major advances in logic and the philosophy of mathematics.

Ryle, Sir Martin (b. Sept. 27, 1918; Brighton, England) British radio astronomer and mathematician. Ryle, the son of a physician, studied at Oxford University. He spent the war with the Telecommunications Research Establishment in Dorset working on radar. After the war he received a fellowship to the Cavendish Laboratory of Cambridge University and in 1948 was appointed lecturer in physics. In 1959 he became the first Cambridge professor of radio astronomy, having been made in 1957 the director of the Mullard Radio Astronomy Observatory in Cambridge. Ryle was appointed Astronomer Royal in 1972 and in 1974 was awarded, jointly with Antony Hewish, the Nobel Prize for physics. He was knighted in 1966. It was mainly due to Ryle and his colleagues that Cambridge, after the war, became one of the leading centers in the world for astronomical research. He realized that one of the first jobs to be done was simply to map the radio sky. He therefore began in 1950 the important series of Cambridge surveys. The first survey used the principle of interferometry and discovered some 50 radio sources. The second survey in 1955 listed nearly 2000 sources, many of which turned out to be spurious. The crucial survey was the third one, the results of which were published in 1959 in the *Third Cambridge Catalogue* (3C). This listed the positions and strengths of 500 sources and has since become the definitive catalog used by all radio astronomers. The use of more sensitive receivers in 1965 enabled the 4C survey to detect sources five times fainter than those in the 3C and covered the whole of the northern sky; 5000 sources were cataloged. Finally, with the opening of two highly sensitive radio telescopes in 1965 and 1971, important areas of the sky are being surveyed in depth: a full survey would take over 2000 years. The two new telescopes, the One Mile telescope and then the Five Kilometer telescope, operate by a technique developed by Ryle and called 'aperture synthesis'. A number of radio dishes are used to give a very large effective aperture, much larger than the aperture of a single dish, and hence produce very considerable resolution of detail in a radio map of an area of the sky. The dishes are mounted along a line, some fixed in position, some moveable, and are used in pairs, at different distances apart, to form interferometers. These telescopes and other equipment have been used by Ryle and his colleagues to investigate pulsars, which were discovered at Cambridge by Antony Hewish and Jocelyn Bell, quasars, radio galaxies, and other radio sources. Ryle quickly appreciated that the distribution of radio sources throughout the universe had cosmological implications and that the number of sources found tended to support the evolutionary bigbang theory rather than the steady-state theory.

S

Schwarzschild, Karl. (b Oct. 9, 1873; Frankfurt am Main, now in West Germany; d. May 11, 1916; Potsdam, now in East Germany). German astronomer Schwarzschild was the son of a prosperous Jewish businessman. His interest in astronomy arose while he was at school and he published two papers on binary orbits by the time he was 16. Following two years at the University of Strasbourg, he went in 1893 to the University of Munich, obtaining his PhD in 1896. He worked at the Kuffner Observatory in Vienna from 1896 to 1899 and after a period of lecturing and writing became in 1901 associate professor, later professor, at the University of Göttingen and director of its observatory. In 1909 he was appointed director of the Astrophysical Observatory in Potsdam. He volunteered for military service in 1914 at the beginning of World War I and was invalided home in 1916 with a rare skin disease from which he died. Schwarzschild's practical skills was demonstrated by the instruments he designed, the measuring techniques he devised, and the observations he made. In the 1890s, while the use of photography for scientific purposes was still in its infancy, he developed methods whereby the apparent magnitude, i.e. observed brightness, of stars could be accurately measured from a photographic plate. At that time stellar magnitudes were usually determined by eye. He was then able to establish the photographic magnitude of 3500 stars brighter than magnitude 7.5 and lying between 0° and 20° above the celestial equator. He also determined the magnitude of the same stars visually, demonstrating that the two methods do not yield indentical results. The difference between the visual and photographic magnitude of a star, measured at a particular wavelength, is known as its color index. Schwarzschild also made major contributions to theoretical astronomy, the subjects including orbital mechanics, the curvature of space, and the surface structure of the Sun. In 1906 he published a paper showing that stars could not just be thought of as a gas held together by its own gravity. Questions of thermodynamics arise, concerning the transfer of heat within the star both by radiation and convection, that need a full mathematical treatment. Einstein's theory of general relativity was published in 1916. While serving in Russia, Schwarzschild wrote two papers on the theory, which were also published in 1916. He gave a solution — the first to be found — of the complex partial defferential equations by which the theory is expressed mathematically and introduced the idea of what is now called the *Schwarzschild radius.* When a star, say, is contracting under the effect of gravity, if it attains a particular radius then the gravitational potential will become infinite. An object will have to travel at the velocity of light to escape from the gravitational field of the star. The value of this radius, the Schwarzschild radius, SR, depends on the mass of the body. If a body reaches a radius less than it SR nothing, including light, will be able to escape from it and it will be what is now known as a 'black hole'. The SR for the Sun is 3 km while its actual radius is 700 000 km. The theoretical study of black holes and the continuing search for them has become an important field in modern astronomy. Schwarzschild's son, Martin, is also a noted astronomer.

Sedgwick, Adam. (b. Mar. 22, 1785; Dent, England; d. Jan. 27, 1873; Cambridge, England). British geologist and mathematician. Sedgwick was the son of the vicar of Dent. He graduated in mathematics from Cambridge University in 1808, was made a fellow in 1810, and was elected to the Woodwardian Chair of Geology in 1818, a post he retained until his death. He was made president of the Geological Society in 1829. In 1831 he began a study of the Paleozoic rocks of Wales, choosing an older region than the Silurian recently discovered by Roderick Murchison. In 1835 he named the oldest fossiliferous strata the Cambrian (after Cambria, the ancient name for Wales). This immediately caused a problem for there was no reliable way to distinguish the Upper Cambrian from the Lower Silurian. Sedgwick formed a close friendship with Murchison. The two made their most significant joint investigation with their identification of the Devonian System from studies in southwest England in 1839. The partnership between Sedgwick and Murchison was broken when Murchison annexed what Sedgwick considered to be his Upper Cambrian into the Silurian. The bitter dispute between the two over these Lower Paleozoic strata was not resolved until after Sedgwick's death, when Charles Lapworth proposed that the strata should form a new system—the Ordovician. Sedgwick's works included *A Synopsis of the Classification of the British Paleozoic Rocks* (1855). In 1841, largely due to Sedgwick, a museum now bearing his name was opened to house the growing geological collection. Sedgwick was, throughout his life, a committed opponent of Darwin's theory of evolution.

Serre, Jean-Pierre (b. Sept. 15, 1926; Bages, France). French mathematician. Serre, one of the outstanding French mathematicians of his generation, studied at the Ecole Normale Supérieure, and has taught at both the University of Nancy and Princeton University. He has been a professor at the Collège de France since 1956, and a member of the French Academy of Sciences since 1976. Serre's mathematical work began with a collaboration with Henri Cartan in which they were able to bring about a decisive reorientation of the theory of a complex variable. What Serre and Cartan did was to reformulate, in a completely original way, the central problems and results of the subject in terms of what is known as cohomology theory. Serre himself also did work of great originality and importance on the homotopy theory of spheres. Homotopy and homology theory have been the area of his greatest mathematical achievements. Before his revolutionary work, the two subjects had been thought of as quite unconnected and unrelated. But Serre was able to show how to associate the homotopy group of a space with the homology groups of suitably constructed auxiliary spaces. By forging this link he succeeded in bringing to bear purely algebraic methods and techniques on the central problems of homotopy theory in a way that was entirely new and immensely fruitful. In recognition of this fundamental work in revolutionizing homotopy theory Serre, aged only 28, was awarded a Field's Medal in 1954.

Shannon, Claude Elwood (b. Gaylord, Mich., USA, 30.4.1916). American pioneer of mathematical communication theory, Shannon was educated at the University of Michigan, and after academic research joined the Bell Telephone laboratories. In 1948 he published a classic paper 'The mathematical theory of communication' in the Bell *System Technical Journal.* Shannon was seeking to give a coherent treatment of all forms of information transmission systems, whatever their physical nature. His paper used such terms as information , redundancy and message, which are now commonplace. Developed further in a series of papers, Shannon's work is fundamental to all modern communication systems. In 1949 he published with Weaver *The Mathematical Theory of Communication* (Urbana).

Shapley, Harlow. (b. Nov. 2, 1885; Nashville, Missouri; d. Oct. 20, 1972; Boulder, Colorado) American astronomer. Shapley came from a farming background. He began his career as a crime reporter on the *Daily Sun* of a small Kansas town when he was 16. He entered the University of Missouri in 907 intending to study journalism but took astronomy instead, gaining his AM in 1911. He then went on a fellowship to Princeton where he studied under Henry Russell and gained his PhI) in 1913. From 1914 to 1921 he was on the staff of the Mount Wilson Observatory in California. Finally Shapley was appointed in 1921 to the directorship of the Harvard College Observatory where he remained until 1952, also serving for the period 1922-56 as Paine Professor of Astronomy. Shaply's early work, under Russell, on eclipsing binaries proved that the group of stars, known as Cepheids, were not binary but were single stars that changed their brightness as they changed their size. Cepheids were thus the first 'pulsating variables' to be discovered, the theory of the pulsation being supplied subsequently by Arthus Eddington. Once at Mount Wilson, Shapley began to study Cepheids in globular clusters, huge spherical groups of closely packed stars. From this stemmed his fundamental work on the size and structure of our Galaxy. In 1915 he was able to make a bold speculation about the galactic structure. Using the relation between the period of Cepheids and their observed brightness, discovered in 1912 by Henrietta Leavitt, he was able to map the relative distances of clusters from us and from each other. To his surprise he found that they were widely and randomly distributed both above and below the plane of the Milky Way and appeared to be concentrated in one smallish area in the direction of the constellation Sagittarius. He argued that such a distribution would make sense if the Galaxy had the shape of a flattened disk with the clusters grouped around the galactic center. This required that the solar system be displaced from its accepted central position by a considerable distance. Thus Shapley had found the general structure of the Galaxy but not its size. Here the Cepheids were of limited use as they could only provide a relative scale. Absolute distances could at that time only be determined for small distances. In order to calibrate his galactic structure Shapely needed to measure the distance of a few Cepheids. He used a statistical method pioneered by Ejnar Hertzsprung in 1913. Since the intrinsic brightness, or luminosity, of stars can be determined once their distance is known, Shapley's measurements allowed him to produce a quantified form of the relationship between Cepheid period and observed brightness, i.e. a period-luminosity relationship. This P-L relationship meant that a measure of the period of any Cepheid would reveal its luminosity and hence its distance and the distance of the stars surrounding it. By 1920 Shapley felt that he had finally cracked the fundamental problem of the scale of the Galaxy. The

Sun, he declared, was some 50 000 light-years from the center of the Galaxy while the diameter of the galactic disk could be perhaps 300 000 light-years. Actually Shapley' calculations were too generous as he was unaware of the interstellar matter that absorbs some of the light from stars and thus affects determinations of stellar brightness. Consequently his figures were later revised to 30 000 light-years for the distance to the galactic center and 100 000 light-years for the diameter. Shapley was however less successful with his work on the scale of the Universe. In 1920 he took part with Heber Curtis in a famous debate organized by the National Academy of Sciences at the Smithsonian in Washington. Using the brightness of novae in the Andromeda nebula, Curts gave an estimate approaching 500 000 light-years for its distance and maintained that it was an independent star system. Shapley, misled by the measurements of Adriaan van Maanen, argued that this distance was far too great and that the Andromeda nebula and the other spiral nebulae lay within the Galaxy. It was left to Edwin Hubble to show, some years later, that Curtis had in fact underestimated rather than overestimated the distance of the Andromeda nebula and that it was in fact a separate star system. Not the least of Shapley's achievements was his development of the Harvard Observatory into one of the major research institutions of the world. He introduced a graduate program and attracted a distinguished and much increased permanent staff. During his time there his interest turned to 'galaxies', as he called them, or 'extragalactic nebulae' in Hubble's terminology. Northern and southern skies were surveyed for galaxies and tens of thousands were recorded. In 1932 he produced a catalog, with Adelaide Ames, of 1249 galaxies, which included over a thousands galaxies brighter than 13th magnitude. In 1937 he published a survey of 36 000 southern galaxies. He also studied the Magellanic Clouds and identified the first two dwarf galaxies, the Fornax and Sculptor systems, which are members of the Local Group of galaxies. Shapley wrote several books on astronomy and left an account of his scientific life in his informal *Through Rugged Ways to the Stars* (1969).

Sheldon Glashow. Sheldon Glashow was born in New York City in 1932 and graduated from the Bronx High School of Science. He earned his undergraduate degree at Cornell in 1954 and received a doctorate in physics from Harvard in 1959. Glashow did postdoctoral research at the Bohr Institute, Cern in Geneva and the California Institute of Technology. He began his faculty career at Berkeley, but left in 1966 to become a physics professor at Harvard. Glashow's work, which ultimately earned him one-third of the 1979 Nobel Prize in physics, was for helping to develop the theory of the socalled "electroweak force," a theory that is said to unify the weak and electromagnetic interactions between elementary particles, and which predicts the neutrino neutral current. Howard Georgi and Glashow have extended the electroweak unification to try to include the strong interaction in it and thus to obtain a grand unification theory (GUT); this effort has not been successful. Glashow also introduced a fourth quark called "charm" to account for a certain discrepancy between the theoretical prediction of the rate of decay of certain strange mesons, called kaons, and the observed rate of decay. The introduction of a charmed quark replaced the Gell-Mann SU(3) symmetry group by the SU(4) group. The evidence for the charmed quark is assumed to be the discovery in 1974 of the J/psi particle by Burton Richter and Samuel C. C. Ting. Glashow's article provides a concise historical overview of particle physics developments that date back to the time of Rutherford. He shows how a multitude of elementary particles can be explained by combinations of up, down , and strange quarks. The charm quark is used to explain in a systematic fashion the proliferation of baryons and mesons. Glashow also explains the functions of gluon particles, "gluons," which bind the quarks inside hadrons and produce the exchange of "colors" among quarks—names given to three new kinds of charge on each quark and designated as "red," "green." and "blue." Although gluons must also have colors, quark theory states that no colored particle can exist in isolation. Particles exist only when their constituent quarks combine so that the resulting color is "white." Any attempt to extract a single "colored" quark fails, because the introduction of the energy into that subatomic system to extract a quark creates only combinations of quarks and antiquarks which are also color-neutral.

Srinivasa Ramanujan. The greatness of Ramanujan as a mathematician can be judged from the fact that his birth centenary was celebrated widely in December 1987. His contribution to the theory of numbers brought him worldwide acclamation. One famous incident of his life is as follows. When he was a child, the teacher was taking the arithmetic class. On the black-board three bananas were drawn. The teacher asked the students: "If we have three bananas and three boys how many bananas will each boy get?" A smart boy in the front row replied: "Each will get one" "Right", said the teacher. While the teacher was explaining the method of division, a boy sitting in one corner asked: "Sir, if no banana is distributed among on one, will every one still get the banana?" "What a silly question to ask?". said all the students with a roar of laughter. The teacher, however, was very much impressed with the question and said: "There is nothing to laugh at about. I will explain what he means to say. He is asking if zero is divided by zero, will the result be one." The boy had asked a question that had taken mathematicians several centuries to answer. Some mathematicians claimed that zero divided by zero was zero, but others claimed it to be unity. The correct answer was found by Indian mathematician Bhaskara who proved that zero divided by zero is equal to infinity. The boy who had asked the astounding question was Srinivasa Ramanujan who later became a great mathematician. This Tamil child's father was a petty clerk in a cloth shop. Because of his poor financial condition, Ramanujan could not get proper education. When he was just 13, he solved the world famous Loney's Trignometry. At the age of 15, he obtained a copy of George Schoobsidge Carr's Synopsis of Elementary Results in Pure and Applied Mathematics. This book had a collection of about 6000 theorems. He verified all these theorems, and on this basis he developed some new theorems. In 1903, he secured a scholarship from the University of Madras. But since he was devoted totally to mathematics, he neglected other subjects in which he failed and consequently the scholarship was withdrawn the following year. His father was shocked. When he found that his son was all the time playing with numbers. He thought Ramanujan had gone mad. To set him right, he forced his son to marry. The girl chosen was eight-year old Janaki. After his marriage, Ramanujan began to look for a job. He had to find money not only for his bread and butter, but for paper as well to do his mathematical calculations on. He started using even scraps of paper he found lying in the streets. Sometimes he would use a red pen to write over blue ink in order to use the same paper twice. He managed to get a clerical job on a monthly salary of Rs. 25. Later, some teachers and educationists interested in mathematics initiated a move to provide Ramanujan with research fellowship. He was granted a fellowship of Rs. 75; a month form University of Madras, though he had no qualifying degree. During these days he wrote a letter to the great mathematician G. H. Hardy of Cambridge University and also sent 120 theorems to him. Hardy and his colleagues realised the greatness of the work. They made arrangements for Ramanujan's passage and stay at Cambridge University. On March 17, 1914 Ramanujan sailed for Britain. Ramanujan found himself a stranger at Cambridge. The cold was hard to bear and being a Brahamin and Vegetarian, he had to cook his own food. However, he continued his research in mathematics. In Ramanujan, Hardy found an unsystematic mathematician who played with numbers. For his work, he was elected Fellow of the Royal Society on February 28,1918. He was the second Indian to receive this distinguished fellowship. In October of the same year, he became the first Indian to be elected Fellow of Trinity College, Cambridge. In algebra, his work is considered to be equal in importance to that of great mathematicians like Eular and Jacobi. While Ramanujan continued his research work in England, tuberculosis was affecting him. He was sent back to India. He had become pale and very weak. Even in these conditions. He continued to play with numbers. He died of tuberculosis on April 26,1920 at Chetpet in Madras. Besides being mathematician, he was an astrologer of repute and a good speaker. He used to give lectures on subjects like "God, and Infinity". Ramanujan award was instituted in order honour him. Ramanujan Institute was also founded in his memory which is functioning even today under the University of Madras.

Simon Newcomb. The foremost American astronomer of his time and one of the most eminent men of science our country has ever developed, was born in Wallace, Nova Scotia, March 12, 1835, of humble parents. His father was a country school teacher and from him the obtained his early education. Apprenticed to a physician at 16 years of age, he ran away from a thankless and profitless job of general factotum and, in 1853, came to the United States. Here he taught school in Maryland in different country towns, finally getting one near Washington where "the Smithsonian Institution library was one of the greatest attractions". There he saw for the first time Laplace's *Mécanique céleste"*, "the greatest treasure that my imagination had ever pictured". He had in his early school days shown fondness for the study of geography and navigation and he now made application for a post in the Coast Survey. In 1857 he found employment as computer at the office of the "Nautical Almanac"—then in Cambridge, in order to be near the technical experts, especially Benjamin Peirce, of Harvard—and at the same time took the course at the Lawrence Scientific School, where he was graduated in 1858 and afterwards remained for three years as a resident graduate. He was then appointed professor of mathematics in the U.S. Navy and was assigned to work in the ill-equipped naval absservatory. Here, as better instruments were installed and he was more and more given a free hand by his superiors, he began the work of determining the error in the right ascension of stars which he believed had crept into the modern observations made at Greenwich, Paris and Washington. This task took him a number of years and in 1869 he found after working up his observations that the error he had suspected was real. This faculty of tireless patience in sifting mathematical details, combined with skill and originality in approaching difficult astronomical problems and a talent for observation, was characteristic of Newcomb. In 1877, being then senior professor of mathematics in the Navy with rank of captain, he was assigned to the charge of the "Nautical Almanac" office which had been moved to Washington in 1866. He says: "The change was one of the happiest of my life. I was now in a position of recognized responsibility, where my recommendations met with respect due to that responsibility, where I could make plans with the assurance of being able to carry them out, and where the countless annoyances of being looked upon as an important factor in work where there was no change of my being such would no longer exist. Practically I had complete control of the work of the office, and was thus, metaphorically speaking, able to work with untied hands." He had soon reformed the office, which he concentrated no one floor of a modern city building, making economies that "went on increasing year by year and every dollar that was saved went into the work of making the tables necessary for the future use of the *Ephemeris*". His program included a consideration and recalculation of all the observations of all the leading observatories made since 1750 on the positions of the sun, moon and planets, as well as on the bright fixed stars, and included as well the compilation of the formulae for the perturbation of the planets by each other. These were tasks the magnitude of which beggars a layman's understanding. "The number of meridian observations of the sun, Mercury, Venus and Mars alone numbered 62,030, and they were contributed by the observatories at Greenwich, Paris, Königsburg, Pulkowa, Cape of Good Hope and elsewhere." It was this task of "bringing the great problem of the solar system well-night to completeness of solution" that constituted Newcomb's life work and on which his reputation rests. It necessitated "an almost complete reconstruction of the theories of the motions of the bodies of the solar system" and "at its foundation the complete revision of the So-called constants of astronomy". His early observations at the Naval Observatory had shown him "that the moon seemed to be falling a little behind her predicted motion". Investigation convinced him that the error had crept into the records prior to 1750 and he found evidence at Pulkowa that it was before 1675. This error, in the time that had since elapsed, had with slight additions amounted to enough to considerably affect all astronomical records. It took him years to compile the necessary data and it was only in 1878 that he published his "Reductions and Discussion of the Moon before 1750", and only on his death bed that he completed "The Motion of the Moon". These are but a few of the herculean tasks undertaken by this marvelous worker and carried to a successful conclusion. One should consult his modest "Reminiscences of an Astronomer" to see how full be made each day of his happy life. He supervised the construction of the great 26-inch telescope for Washington, "the success of which excited such interest the world over as to give new impetus to the construction of such instruments," resulting in the still larger lenses at Pulkowa, Lick and Yerkes observatories, concerning all of which he was much consulted. In 1884 Newcomb was invited to become professor or mathematics and astronomy at Johns Hopkins University, but he never resided in Baltimore. He resigned in 1893, returned to the chair in 1898 and was made professor emeritus in 1900. He was retired for age from the "Nautical Almanac" office in 1897. He was for many years editor of the *American Journal of Mathematics* and made more than 300 important scientific contributions to astronomy, most of them published in the "Astronomical Papers of the *American Emphemeris*". He wrote a number of popular works on astronomy and mathematics—his "Popular Astronomy" (1878) has been translated into half a dozen languages—and one or two more abstruse, besides a novel, several financial and economics works, an autobiography and articles in many magazines and encyclopedias. His versatility was almost as great as his application in this favorite field. Simon Newcomb's work received the recognition of honorary degrees from sixteen universities, honorary membership in the great academies of science and fellowship in all the learned societies, decorations and medals galore. He was the first American since Franklin to become an associate of the French Institute. The French government made him an officer of the Legion of Honcr in 1896 and a commander in 1907. Simon Newcomb died on July 11,1909 and was given a military funeral, having been accorded the rank or rear admiral by Congress in 1906.

Smith, Henry John (b. Nov. 2, 1826; Dublin; d. Feb. 9, 1883; Oxford, England). British mathematician. Smith studied at Oxford and had a great interest in classics—it was only after a good deal of hesitation that he chose mathematics as a profession instead. He remained in

Oxford in various capacities for most of the rest of life. In 1860 he became Savilian Professor of Geometry there. Smith's main work was in number theory and his greatest contribution was his development of a general theory of *n* indeterminates, which enabled him to establish results about the possibility of expressing positive integers as sums of five and seven squares. This achievement ought to have won Smith the prestigious prize offered in 1882 by the French Academy for their mathematical competition. However Smith, a notably unambitious man, did not enter and the prize was in fact given to Hermann Minkowski. When it was discovered that Minkowski had made use of crucial results published by Smith, the French Academy hastened to transfer the prize to Smith, but as he had unfortunately died in the meantime his fame was only posthumous. Smith also worked on the theory of elliptic functions.

Snell, Willebrord van Roijen (b. 1591; Leiden, Netherlands; d. Oct. 30, 1626; Leiden) Dutch mathematician and physicist. Snell's initial training was in mathematics and came from his father who taught at Leiden University. He traveled widely in Europe visiting Paris, Würzburg, and Prague, and among the celebrated scientists he met were Johannes Kepler and Tycho Brahe. Once he had returned to Leiden, Snell published a number of editions of classical mathematical texts. On the death of his father (1613) Snell succeeded him as professor of mathematics at the university. He was involved in practical work in geodesy and took part in an attempt to measure the length of the meridian. In this project he was one of the first to see the full usefulness of triangulation and published his method of measuring the Earth in his *Eratosthenes Batavus* (1617; The Dutch

Eratosthenes). In 1621 Snell discovered his famous law of refraction, based on a constant known as the refractive index, after much practical experimental work in optics. Snell did not however publish his discovery and the law first reached print in 1638 in a work by Descartes. However, Descartes had arrived at the law in a totally different way from Snell, and made no use of practical observation.

Stephen W. Hawking. Stephen William Hawking was born in 1942 in Oxford, England, and was educated at both Oxford and Cambridge, receiving his Ph.D. from the latter institution. After working in various positions at Cambridge, he was appointed to the chair of gravitational physics there in 1977. He has concentrated his efforts on devising an alternative to Einstein's treatment of gravity in his general theory because "it treats the gravitational field in a purely classical manner when all other observed fields seem to be quantized. Hawking's objection was based on the general theory's prediction of singularities which it could not meaningfully describe. As a result, Hawking has searched for a theory of quantum gravity that can avoid the problems of the point singularity. Although he has made some progress in this endeavor, existing quantum gravity theories are still riddled with a number of theoretical defects. Hawking has fought a crippling nervous disorder to become one of the most eminent physicists in the world. Although he has been confined to a wheelchair and can no longer write or calculate directly or even speak clearly, he continues to propose a staggering array of new physical ideas as well as to make complex calculations and to formulate detailed mathematical proofs. The talented science writer Isaac Asimov has written that "[Hawking] has been reduced to virtual immobility and helplessness but trapped within the dying body is a brilliant mind and an apparently indomitable spirit."* Hawking is especially well known for his suggestion that the Universe is populated by vast numbers of particle-sized black holes which were formed by the explosive compression of matter during the early expansionary phase of the Universe. Like ordinary black holes, these primordial "mini" or Hawking black holes distort space-time but on a much more localized scale. The event horizon of a one-billion-ton primordial black hole has a diameter of only about one ten-trillionth of a centimeter. Yet due to the smallness of their masses, Hawking black holes may emit particles. The uncertainty principle holds that the farther we peer into the subatomic world, the less clearly we are able to distinguish the features of individual particles such as their positions or energies. This uncertainty in the quantum world extends to the relationship between space and time. At any single moment we cannot be certain of the quantity of particles in a specific area of space. As a result, particle-antiparticle pairs can appear and disappear anywhere in space at any time. These particle pairs do not magically appear from nothing but are holdovers from the creation of equal numbers of particles and antiparticles during the big bang. The lives of these particles are very short An electron-antielectron pair, for example, appears and disappears within one billionth-trillionth of a second. Although more massive couplings such as proton-antiproton pairs also appear, the inverse nature of the uncertainty principle is such that the smaller particles exist for proportionately smaller periods of time. Since these particles have such short lives they are said to exist "virtually" and are referred to as virtual pairs. When one of these particle pairs appears next to a

Hawking black hole, one of the virtual-pair particles often finds itself inside the black hole. This position causes the virtual pair to separate, freeing the particle outside the black hole to roam freely throughout the Universe. An outside observer concludes that the primordial black hole has emitted a particle because the detachment transforms the virtual particle into an actual particle. As the energy from the emission of the separated particle can come only from the gravitational field of the Hawking black hole, It will lose mass over time as the energy of its gravitational field is depleted until it evaporates completely. This phenomenon occurs mostly for microscopic black holes, however, because it is unlikely that during the instant of particle separation that only one particle of a virtual pair would find itself outside the gravitational field of an ordinary black hole. The tendency of fleeing particles to steal the energy fields of Hawking black holes increases as the mass of the black hole decreases. There is also an inverse relationship between the size of the black hole and its temperature. As it radiates energy, its mass decreased and its temperature increases—thereby increasing the rate of energy loss. In short, the smaller the black hole, the more probable its evaporation. For a Hawking black hole to have survived until the present day, it must have originally possessed a mass of at least several billion tons. Smaller black holes would have evaporated by this time. The time needed for a black hole with the mass of our sun to evaporate is vast—about 10^{66} years. If we live in an oscillating Universe that expands and contracts every hundred billion years (10^{11}) our Universe would go through about 10^{56} additional hundred-billion-year cycles of expansion and contraction before the tensolar-mass black hole had completely evaporated. Our selection contains Hawking's examination of the quantum mechanics of black holes.

Stevin, Simon (b. 1548; Bruges, now in Belgium; d. May 1620; The Hague, Netherlands). Flemish mathematician and engineer. Steven was also known as Stevinus, the Latinized form of his name. He worked for a time as a clerk in Antwerp, eventually working his way up to become quartermaster of the army under Prince Maurice of Nassau. While in this post he devised a system of sluices, which could flood the land as a defense should Holland be attacked. Stevin was a versatile man who contributed to several areas of science. Mathematics owes to him the introduction of the decimal system of notating fractions. This system was perfected when John Napier invented the decimal point. Stevin helped to popularize the practice of writing scientific works in modern languages (in his case Dutch) rather than Latin, which for so long had been the traditional European language of learning. However such was the hold of the old ways that Willebrord Snell thought it was worthwhile to translate some of Stevin's work into Latin. To hydrostatics he contributed the discovery that the shape of a vessel containing liquid is irrelevant to the pressure that liquid exerts. He also did some important experimental work in statics and in the study of the Earth's magnetism.

Stokes, Sir George Gabriel (b. Aug. 13, 1819; Skreen, now in the Republic of lreland; d. Feb. 1, 1903; Cambridge, England) British mathematician and physicist. Stokes studied at Cambridge and remained there throughout his life. In 1849 he became Lucasian Professor of Mathematics, but he found it necessary to supplement his slender income from this post by teaching at the

Government School of Mines in London. He held his Cambridge chair until his death aged 84. He was MP for the university and among his many honors were a baronetcy conferred on him in 1889. Stokes was equally interested in the theoretical and experimental sides of physics and did important work in a wide area of fields, including hydrodynamics, elasticity, and the diffraction of light. In hydrodynamics he derived the formula now known as *Stokes's law*, giving the force resisting motion of a spherical body through a viscous fluid. Among Stokes's other fields of study was fluorescence—one of his experimental discoveries was the transparency of quartz to ultraviolet light. He was also much interested in the then influential concept of the ether as an explanation of the propagation of light. Stokes became aware of some inherent difficulties with the concept, but rather than rejecting the whole idea of an ether he tried to explain these problems away by using work he had done on elastic solids, though naturally enough problems arose with his own ideas. Stokes was perceptive in his views of other physicists' work. For example, he was among the first to appreciate the importance of the work of James Joule and to see the true meaning of the spectral lines discovered by Joseph von Fraunhofer.

Sylvester, James Joseph (b. Sept. 3, 1814; London; d. Mar. 15, 1897; London). British mathematician. Sylvester studied mathematics at Cambridge but was not granted his BA degree since he was a practicing Jew. The relevant statute was later revoked and Sylvester was granted both his BA and MA in 1817. He was widely read in a number of languages and was a keen amateur musician and a prolific poet. Feeling unable to keep an academic post as a mathematician Sylvester worked first in an insurance company and later as a lawyer. In 1876 he went to America to become the first professor of mathematics at Johns Hopkins University. He became the first editor of the *American Journal of Mathematics*, and did much to develop mathematics in America. He returned to England in 1883 to become Savilian Professor of Geometry at Oxford University. Sylvester's best mathematical work was in the theory of invariants and number theory. With his life-long friend the British mathematician Arthur Cayley, he was one of the creators of the theory of algebraic invariants, which proved to be of great importance for mathematical physics.

T

Tarski, Alfred (b. Warsaw, Russian Poland [now Poland], 14.1.1902). Polish logician and mathematician, a brilliant member of the distinguished group of logicians working in Poland in the interwar years. Excluded, because Jewish, from a faculty university post, Tarski taught concurrently in a high school and at Warsaw university until 1939. Since 1942 he has been based at the University of California at Berkeley. Productive over a wide range of logical and mathematical topics, he made pioneering studies of the 'equivalential calculus' and refined consideration of the notions of the consistency and completeness of formal logical theories. A most notable achievements was his monograph of 1933 ['The concept of truth in formalized languages'] (tr. J. H. Woodger in *Logic, Semantics, Metamathematics,* Oxford, 1956). This has been the starting-point for all logically serious discussions of the subject, ever since. In it Tarski rehabilitates the classical correspondence theory of truth in contemporary logical guise. Since WW2 Tarski has continued to work on a wide variety of logical and mathematical topics, usually of a rather specific and technical kind. More generally his work has served to legitimate semantic discourse about the relations between language and the world, explicitly proscribed by

Wittgenstein although practised by him. Quine's critique of the analytic synthetic distinction was foreshadowed by Tarski, and he also had much influence on Carnap and Popper. K. Popper, *Objective Knowledge* (Oxford, 1974).

Tarski, Alfred (b. Jan. 14, 1902; Warsaw). Polish American mathematician and logician. Tarski served as a professor at the University of Warsaw (1925-39). In 1942 he joined the staff at the University of California and became professor there in 1949 and research professor at the Miller Institute (1959-60). Tarski has worked on set theory and algebra and is noted as one of the pioneers in the study of formalized logical systems as purely algebraic structures. He emphasized the difference between the metalanguage, used to talk about these structures, and the formal language whose syntax formed the system being studied. His famous paper *The Concept of Truth in Formalized Languages* (1935) was one of the foundation stones of model theory, and has had a profound influence both in logic and the philosophy of language.

Tartaglia, Niccolò (b. 1500; Brescia, Italy; d. Dec. 15, 1557; Venice, Italy). Italian mathematician, topographer, and military scientist. Tartaglia's true name was Fontana but as a boy he suffered a saber would to his face during the French sack of Brescia (1512), which left him with a speech defect. He adopted the nickname Tartaglia (Stammerer) as a result. Tartaglia began his studies as a promising mathematician, but his interests soon became very wide ranging. He held various posts, including school teacher, before he eventually became a professor of mathematics in Venice where he stayed. Tartaglia is remembered chiefly for his work on solving the general cubic equation. He discovered a method in 1535 but did not publish it. Incautiously he revealed his new method to his friend the mathematician Girolamo Cardano who published it in *Ars magna* (1545; The Great Skill). This, not surprisingly, was the end of their friendship and led to a violent controversy. Tartaglia eventually lost the quarrel and with it his post as lecturer at Brescia in 1548. Tartaglia's other chief mathematical interests were in arithmetic and geometry. The pattern now known as 'Pascal's triangle' appeared in a work of Tartaglia's. His geometrical work centered on problems connected with the tetrahedron and he helped further the diffusion of classical mathematics by making the first translation of Euclid's *Elements* into a modern European language. His chief published work was the three-volume *Trattato di numeri et misure* (1556-60; Treatise on Numbers and Measures), an encyclopedic work on elementary mathematics. Apart from these mathematical activities Tartaglia made notable innovations in topography and the military uses of science, such as ballistics.

Taylor, Brook (1685-1731). British analyst, geometer, painter, and philosopher, who pioneered the infinitesimal calculus and wrote two works on perspective. Because he did not publish his results, some were claimed by Johann Bernoulli, and the importance of Taylor's Theorem was only recognized some 60 years later, by Lagrange. He became a Fellow of the Royal Society, and sat on the committee that adjudicated between the claims of Newton and Leibniz to have first invented infinitesimal calculus. Taylor studied at Cambridge University, and was secretary to the Royal Society during the period 1714-18. He made important contributions to the development of the differential calculus in his *Methodus incrementorum directa et inversa* (1715; Direct and Indirect Methods of Incrementation). This contained the formula known as *Taylor's Theorem,* which was recognized by Joseph Lagrange in 1772 as being the principle of differential calculus. The *Methodus* also contributed to the calculus of finite differences, which Taylor applied to the mathematical theory of vibrating strings. Outside mathematics Taylor was an accomplished artist and led him to an interest in the theory of perspective, publishing his work on this subject in *Linear Perspective* (1715). *Taylor expansions* are named for him.

Taylor, Geoffrey Ingram (b. London, UK, 7.3.1886; d. Cambridge, 27.6.1975). British physicist and applied mathematician. One of the most prolific and wide ranging scientists of the century, Taylor was educated at Cambridge, where, apart from work at the Royal Aircraft Establishment, Farnborough, during WWI, he spent his professional life and where he was appointed Yarrow research professor in 1923. He was the classical physicist *par excellence* and was outstanding for a very wide variety of studies, especially in the mechanics of fluids and solids. Of particular importance was his work on turbulent motion in fluids. This he applied to problems in meteorology and oceanography and even to the mechanics of swimming of small creatures. During WW2 he made important contributions to the knowledge of shock waves in explosions. He was one of the first people to make a systematic study of the plastic deformation of crystalline materials and he invented

(1934) the concept of a 'dislocation'. This is a form of atomic misarrangement in a crystal which, if present, enables the crystal to deform at a stress may be 1000 times less than that of an ideally perfect crystal. More recent work has shown that, unless very carefully prepared, all crystals contain large numbers of these dislocations and it is the way in which they move and interact which controls their strength and the way they deform.

Thales (b. *c*. 625 BC; Miletus; now in Turkey; d. *c*. 547 BC). Greek philosopher, geometer, and astronomer. It is with Thales that physics, geometry, astronomy, and philosophy have long been thought to begin. Little is however known of the first supposed identifiable 'scientist' apart from a number of anecdotes that clearly originate in folklore. Thus, traditionally, he is supposed to have acquired his learning from Egypt, an implausible claim when the modest mathematical skills of sixth-century Egypt are contrasted with the supposed achievements of Thales. To him is even attributed a proof of the proposition that the circle is divided into two equal parts by its diameter, a theorem not to be found in Euclid some 300 years later. It is reported by the historian. Herodotus that Thales predicted a solar eclipse in 585 BC. The remaining claim for Thales rests on his introduction of naturalistic explanations of physical phenomena in opposition to the customary understanding of nature in terms of the behaviors of the gods. Hence the importance of his claim that everything is water, perhaps the first recorded general physical principle in history.

Thom, René Frédéric (b. Montbéliard. Doubs, France, 2.9.1923). French mathematician. In the 1950s he developed a powerful geometric theory of differentiable manifolds, called cobordism theory, for which he was awarded the Fields medal in 1958. This is a deep cohomology theory, i.e. it turns important geometric problems into algebraic ones to seek solutions. A major consequent success of the theory was J. Milnor's surprising discovery in 1956 that spheres of dimension 7 or more can possess inequivalent differentiable structures; another was S. Smale's affirmative solution of the J. H. Pioncare conjecture in dimensions greater than 5. During the 1960s Thom developed catastrophe theory which he (and Arnold in Moscow) used to illuminate singularities of differentiable mappings. Since then the theory has found many applications, some in biology as Thom hoped, though nones as yet of the power he would wish, and some much more controversial from which he has disassociated himself. He has taught at Grenoble, Strasbourg and, since 1963, at L'Institut des Hautes Etudes Scientifiques at Bures-sur-Yvette where he is professor of mathematics.

Turkey, John Wilder (b. New Bedford, Mass., USA, 16.6.1915). American statistician, of Princeton and of Bell Telephone laboratories. In the 1950s he developed the jack-knife technique, sacrificing precision to universality. With a few well-defined exceptions, any statistic, especially one whose sampling behaviour is little-known, can be treated to produce an improved estimate together with a variance estimate and viable confidence intervals. In the 1960s Tukey produced the system of exploratory data analysis by which most types of data can be reduced to a smoothed form, and exceptional or erroneous items identified. This suggests hypotheses for testing, and is most useful if different sections of the data need differing models (see his *Exploratory Data Analysis,* NY, 1977).

Turing, Alan Mathison (b. London, UK, 23.6.1912; d. Wilmslow, Ches., 7.6.1954). British mathematician and pioneer in computer theory. Turing graduated from King's College, Cambridge, where he was elected to a fellowship in 1935. In 1937, working at Princeton, he published a paper 'On computable numbers, with an application to the *Entscheidungsproblem*', and developed the concept of a theoretical computer, the Turing Machine. This concept, which he is believed to have derived from the string of operation cards used to determine the sequence of operations in Babbage's analytical engines, is fundamental to the modern theory of computability. In 1938 Turing returned to the UK, and during WW2 he worked on code-cracking at Bletchley Park. In 1945 he joined the National Physical Laboratory at Teddington to work on the construction, design and use of a large automatic computer to which he gave the name of ACE (Automatic Computing Engine). In 1948 he accepted a readership at Manchester University and became assistant director of MADAM (Manchester Automatic Digital Machine), the computer with the largest memory capacity in the world at that time. His efforts in the construction of such early computers and the development of early programming techniques were of major importance. Turing was much interested in the question of whether machines can think. One of his suggestions was that machine thought would more closely resemble human thought if a randcm element could be introduced, such as a roulette wheel. He proposed as a criterion of a machine's intelligence the question of whether anyone can tell if his or her queries are being answered by a machine or a person at the far end of a data-link. Such criteria ignore the physical—particularly the quantum-mechanical—basis of living systems, and are philosophically profoundly unsatisfying. However, from a mathematical point of view, Turing's criterion has proved valuable. In 1952 Turing published the first part of his study of morphogenesis, a theoretical attempt to show how a uniform and symmetrical structure could develop into an asymmetric structure with a definite pattern as a result of diffusion. He died, however, before his work could be finished. S. Turing. *Alan M. Turing* (Cambridge, 1959).

Turing, Alan Mathison (b. June 23, 1912; London; d. June 7, 1954; Wilmslow, England). British mathematician. Turing studied at Cambridge University, where he gained a fellowship in 1935. From 1936 to 1938 he worked at Princeton. He was at the National Physical Laboratory from 1945 to 1948 and subsequently became reader in mathematics at Manchester University. His most important work consisted in giving a precise mathematical characterization of the intuitive concept of computability. Turing developed a precise formal characterization of an idealized computer—the *Turing machine*—and equated the formal notion of effective decidability with computability by such a machine. Turing was thus able to show that a number of important mathematical problems could have no effective decision procedure. In 1937 he showed the undecidability of first-order predicate calculus and in 1950 that of the word problem for semigroups with cancellation. He was able to put his theoretical work on computability into practice during his time at the National Physical Laboratory when the ACE computer was built under his supervision, and later when he became assistant director of the Manchester automatic digital machine. Turing had the misfortune to be homosexual at a time when this was still a criminal offense in England. He committed suicide as a direct result of a prosecution for alleged indecency.

V

Van Vleck, John Hasbrouck (b. Mar. 13, 1899; Middletown, Connecticut; d. Oct. 27, 1980). American physicist and mathematician. Van Vleck was educated at the University of Wisconsin, where he graduated in 1920. Moving to Harvard University he gained his master's degree (1921) and his doctorate (1922), and stayed for a further year as an instructor. From Harvard he went to the University of Minnesota, where he became a full professor in 1927, returned to Wisconsin in 1928, and then went back to Harvard in 1934. Van Vleck is regarded as the founder of the modern quantum mechanical theory of magnetism. His earliest papers were on the old quantum theory, but with the advent of wave mechanics pioneered by Paul Dirac he began to look at the implications for magnetism in particular. In the field of paramagnetism, he introduced the concept of temperature-independent susceptibility, now known as *Van Vleck paramagnetism*. He also made calculations of molecular structure that shed new light on chemical bonding, and he developed ways of describing the behavior of an atom or an ion in a crystal. Another important contribution of Van Vleck was to point out the importance of electron correlation—the interaction between the motion of electrons—for the appearance of local magnetic moments in metals. During World War

II, Van Vleck worked on radar, and showed that at about 1.25-centimeter wavelength water molecules in the atmosphere would lead to troublesome absorption, and that at 0.5-centimeter wavelength there would be a similar absorption by oxygen molecules. This was to have important consequences not just for military (and civil) radar systems, but later for the new science of radioastronomy. In 1977, together with Nevil Mott and Philip Anderson, he shared the Nobel Prize for physics for "fundamental theoretical investigations of the electronic structure of magnetic and disordered systems." (Anderson was once a student of Van Vleck's at Harvard.) Van Vleck's work on electron correlation was mentioned specifically for the central role it played in the later development of the laser.

Vernier, Pierre (b. *c.* 1580; Ornans, France; d. Sept. 14, 1637; Ornans). French mathematician. Vernier was educated by his father who was a scientist, and became interested in scientific instruments. He was employed as an official with the government of Spain and then held various offices under the French government. In 1631 Vernier invented the caliper named for him, an instrument for taking very precise measurements. The principle of the vernier scale is described in his book *La Construction, l'usage, et les propriétéz du quadrant nouveau de mathématique* (1631; The Construction, Uses, and Properties of a New Mathematical Quadrant), which also contained some of the earliest tables of trigonometric functions.

Victor F. Weisskopf. Victor Weisskopf had made substantial contributions to quantum electrodynamics, nuclear physics, and elementary particle physics. He is presently Institute Professor Emeritus of the Department of Physics at the Massachusetts Institute of Technology. He was born in Vienna in 1908 and received his Ph.D. at the University of Göttingen in 1934. After working as a research associate at the University of Copenhagen and at the Institute of Technology at Zurich, Weisskopf came to the United States in 1937 and worked as assistant professor at the University of Rochester until joining the Manhattan Project at Los Alamos in 1943. He was appointed to join the faculty at M.I.T. in 1946 and served as director general of CERN in Geneva from 1961 to 1965 before returning to M.I.T. Weisskopf did his earliest work in quantum electrodynamics with Eugene Wigner, studying the emission and absorption of radiation by electrons. In 1934 he and Wolfgang Pauli hypothesized the existence of bosons—charged fundamental particles without spin. It was only twelve years later that their theory was verified by the discovery of mesons. In 1937 Weisskopf introduced the concept of temperature and evaporation into nuclear physics, which essentially applied thermodynamic principles to subatomic systems and accurately described many nuclear reactions. The so-called evaporation occurs when a proton or neutron is fired into a nucleus and the energy of its impact is distributed among the constituent particles, thereby causing the nucleus to heat up and subsequently emit some particles. Weisskopf also introduced the so-called optical model of the nucleus, which drew upon the work by Maria Goeppert Mayer and J.H.D. Jensen on the shell structures of atoms, and helped to clarify the theory of nuclear reactions. Weisskopf extensive background as well as his gift for explaining even the most difficult concepts of particle physics or cosmology makes him an ideal choice for providing a general introduction to recent theoretical developments in physics that link cosmology with particle physics. His is a varied approach, which makes for a succinct yet detailed article that touches upon nearly every area in modern cosmology and is as grand in its sweep as it is economical in its language.

Viete, Francois (Franciscus Vieta) (b. 1540; Fontenay-le-Comte, France; d. Dec. 13, 1603; Paris). French mathematician. Viéte is also known by the Latinized form of his name, Franciscus Vieta. He was educated at Poitiers where he studied law and for a time he practiced as a lawyer. He was a member of the *parlement* of Brittany but because of his Huguenot sympathies he was forced to flee during the persecution of the Huguenots. On Henry IV's accession however he was able to hold further offices and became a privy councillor to the king. He put his mathematical abilities to practical use in deciphering the code used by Spanish diplomats. Viéte's chief work was in algebra. He made a number of innovations in the use of symbolism and several technical terms still in use (e.g. coefficient) were introduced by him. His work is important because of his tendency to solve problems by algebraic rather than geometric methods. By bringing algebraic techniques to bear on them Viéte was able to solve a number of geometrical problems. A particularly longstanding problem—formulated by the Greek geometer Apollonius of Perga—namely how to construct a circle that touches three given circles, was solved in this way by Viéte. Viéte's major work is contained in his treatise *In artem analyticem isagoge* (1591; Introduction to the Analytical Arts) and among other advances in algebra that it contains are new and improved methods for solving cubic equations. Among these are techniques that make use of trigonometric methods. Viéte also developed methods of approximating the solutions to equations.

Vinogradov, Ivan Matreevich (b. Sept. 14, 1891; Milolyub, now Velikiye Luki in the Soviet Union). Soviet mathematician. Vinogradov held a number of posts at various institutions in the Soviet Union. Initially he taught at the University of Perm (1918-20) until appointed professor of mathematics at the Leningrad Polytechnic Institute. In 1925 he became a professor at Leningrad State University and in 1932, chairman of the National Committee of Soviet Mathematicians of the Soviet Academy of Sciences. From 1934 he has been professor of mathematics at Moscow State University. Vinogradov has been preeminent in the field of analytical number theory; i.e. the study of problems posed in purely number-theoretic terms by means of the techniques of analysis. He has published several books, chiefly on various aspects of number theory, which include *A New Method in the Analytical Theory of Numbers* (1937) and *The Method of Trigonometric Sums in the Theory of Numbers* (1947). One of his most outstanding achievements was his solution in 1937 of a problem of Christian Goldbach's. (Not to be confused with Goldbach's famous conjecture about even numbers and primes, which is still undecided.) Vinogradov showed that every sufficiently large odd natural number is a sum of three primes.

Viviani, Vincenzo (b. Apr. 5, 1622; Florence, now in Italy; d. Sept. 22, 1703; Florence). Italian mathematician. Viviani was an associate and pupil of Galileo, although his chief interest was in mathematics rather than in physics. After the condemnation of Galileo's ideas by the Catholic Church it was unsafe for Viviani to pursue his work on Galileo's mathematics. Accordingly Viviani devoted himself to the thorough study of the mathematics of the Greeks, in particular their geometry, and in this field of work he achieved wide fame. In 1696 he was elected a fellow of the Royal Society of London. Viviani was particularly interested in trying to reconstruct lost sections of works by ancient Greek mathematicians, such as the missing fifth book of Apollonius's *Conics*. He also published Italian translations of the works of classical mathematicians including Euclid and Archimedes. He was an associate of the physicist Evangelista Torricelli and collaborated with him in producing a barometer.

Von Neumann, Johann (b. Budapest, Austria-Hungary [now Hungary], 3.12.1903; naturalized American citizen, 1937; d. Washington, DC. USA, 8.2.1957). Hungarian/American mathematician. Von Neumann was an infant prodigy who went on to qualify in chemistry at Berlin University, in chemical engineering at the Technische Hochschule in Zürich, and in mathematics at Budapest University. His definition of ordinal numbers, published at the age of 20, has been universally adopted. Von Neumann matured rapidly as a mathematician, working at the universities of Berlin (1926-9) and Hamburg (1929-30), and then moving to Princeton (1931), first to the university, then to the Institute for Advanced Study (1933) where he spent the rest of his life. Up to WW2 the bulk of Von Neumann's work (150 papers in all) was in pure mathematics. In 1932 he published a formulation and proof of the ergodic hypothesis of statistical mathematics and a book on quantum mechanics, *Mathematische Grundlagen der Quentenmechanik* [The mathematical bases of quantum mechanics], which is still a standard text. In the late 1930s, in collaboration with F. J. Murray, he published an important work on 'rings of operators' (since called Von Neumann algebras) which proved to be of value in quantum theory. During WW2 Von Neumann worked on the theory of the atomic bomb, and this directed his attention to problems of computation. In computer theory he did pioneering work on logical design; the problems of securing reliable answers from a machine with unreliable components; the function of memory; matching imitation of randomness; and the problem of constructing automata capable of reproducing their own kind. There is doubt the origin of the concept of a stored-program computer, but no doubt that Von Neumann produced its classical formulation. *Bulletin of the American Mathematical Society,* 64 (May 1958) was devoted to Von Neumann's life and work. In economics Von Neumann became universally known for two path-breaking contributions, both mathematical and conceptual. One was his multi-sector model of economic growth, an English version of which appeared in the *Review of Economic Studies* (1945-6). This was concerned with the formal properties of efficient growth paths, i.e paths which maximize the economy's rate of growth over some very long period. In Von Neumann's economic system commodities are used wholly as inputs into given technological processes to produce increased quantities of the same commodities. Even with these highly abstract assumptions the delineation of possible growth paths and the exhaustive analysis of their properties was a major intellectual achievement. The mathematical basis of Von Neumann's model—a set of linear simultaneous equations—is the same as that used in input-output analysis (see Leontief), activity analysis and linear programming. His other legacy in economics is the theory of games, which he and Oskar Morgenstern created in their monumental *Theory of Games and Economic Behavior* (NY, 1944). This studies the problems of rational behavior in situations of interdependent decision making, i.e. where one's own course of action depends, or should depend, on what one's competitor or colleague is likely to do (and vice versa). Concepts (such as 'zero-sum' and 'prisoner's dilemma') devised by game theorists have gradually found their way into ordinary (or at least semiordinary) discourse. R. D. Luce & H. Raiffa, *Games and Decisions* (NY, 1957); J. Vanek, *Maximal Economic Growth* (Ithaca, NY, 1968).

Von Neumann, John (originally Johann) (b. Dec. 28, 1903; Budapest; d. Feb, 8, 1957; Washington DC). Hungarian-American mathematician. Von Neumann studied at the University of Berlin, the Berlin Institute of Technology, and the University of Budapest from where he obtained his doctorate in 1926. He was *Privatdozent* at Berlin (1927-29) and taught at Hamburg (1929-30). He left Europe in 1930 to work in Princeton, first at the University and later at the Institute for Advanced Study. From 1943 he was a consultant on the atomic-bomb project. Von Neumann may have been one of the last people able to span the fields of pure and applied mathematics. His first work was in set theory (the subject of his doctoral thesis). Here he improved the axiomatization given by

Earnst Zermelo and Abraham Fraenkel. In 1928 he published his first paper in the field for which he is best known, the mathematical theory of games. This work culminated in 1944 with the publication of *The Theory of Games and Economic Behavior,* which von Neumann had coauthored with Oskar Moregenstern. Not all the results in this work were novel, but it was the first time the field had been treated in such a large-scale and systematic way. Apart from the theory of games von Neumann did important work in the theory of operators. Dissatisfied with the resources then available for solving the complex computational problems that arose in hydrodynamics, von Neumann developed a broad knowledge of the design of computers and with his interest in the general theory of automata became one of the founders of a whole new discipline. He was much interested in the general role of science and technology in society and this led to his becoming increasingly involved in high-level government scientific committees. Von Neumann died at the early age of 54 from cancer.

W

Wallis, John (b. Nov. 23, 1616; Ashford, England; d. Oct. 28, 1703; Oxford., England). English mathematician and theologian. Wallis was educated at Cambridge University (1632-40), obtaining his MA in 1640. His early training was in theology and it was as a theologian that he first made his name. He took holy orders and eventually became bishop of Winchester. He moved to London in 1645 where he became seriously interested in mathematics and in 1649 he was appointed to the Savilian Chair in Geometry at Oxford University. Wallis's most celebrated mathematical work is contained in his treatise the *Arithmetica infinitorum* (1655; The Arithmetic of Infinitesimals). In this work he gave his famous infinite series expression for π. Generally the treatise took the development of 17th century mathematics a significant step nearer Newton's creation of the infinitesimal calculus. Wallis was one of the first mathematicians to introduce the functional mode of thinking, which was to be of such importance in Newton's work. He also did notable work on conic sections and published a treatise on them, *Tractatus de sectionibis conicis* (1659; Tract on Conic Sections), which developed the subject in an ingeniously novel fashion. His writings were certainly read by Newton and are known to have made a considerable impact on him. Before Newton, Wallis was probably one of the most influential of English mathematicians. Wallis wrote a substantial history of mathematics. His other interests included music and the study of language. He was active in the weekly scientific meetings that eventually led to the foundation of the Royal Society in 1662. During the English Civil War he was a Parliamentarian and put his mathematical talents to use in decoding enciphered letter.

Weaver, Warren (b. Reedsberg, Wis., USA, 17.7.1894; d. New Milford, Conn., 28.11.1978). American mathematician noted for his contribution to communication theory. Building very much upon the work of Shannon, he analysed the problem of communication into three parts. The first is the technical problem, the province of information theorists like Shannon; to what degree can one expect the original signal to survive the process of sending intact, and how is it likely to mutate? The second is the semantic problem, one amenable to linguistics; given the signal, perhaps muted, will it be understood as the sender intended (for instance, words may be clear and still convey nothing)? The third is the effectiveness problem, one mainly of social psychology; does the signal, as understood, affect the receivers' conduct as intended (for instance, is it remembered)? Limiting results at the technical level have implications at the second and third. (See his book written with C. E. Shannon, *The Mathematical Theory of Communication,* Urbana, 1949.)

Weierstrass, Karl Theodor Wilhelm (1815-97). German analyst, who contributed especially to the theories of complex variables, power series, elliptic functions, continuity, quadratic forms, and the calculus of variations. He was sent to Bonn University to study law, but left without a degree after four years of drinking and fencing; he then trained as a mathematics teacher and taught for 14 years. During this time, without contact with the mathematical world, he developed an entirely original rigorous approach to analysis that enabled him to describe continuous but nowhere differentiable functions and so completely undermine the intuitive approach to these concepts. After the appearance of a monograph developing the work of Abel on theory of functions, he was awarded an honorary doctorate and was found an academic post; although he did not write much more, his extremely influential lectures were published. Weierstrass spent four years at the University of Bonn studying law to please his father. After abandoning law he trained as a school teacher and spent nearly 15 years teaching at elementary schools in obscure German villages. However, he found time to combine his mathematical researches with his school teaching and in 1854 he attracted considerable favorable attention with a memoir on Abelian functions, which he published in Crelle's journal. The fame this work brought him resulted in his obtaining a post as professor of mathematics at the Royal Polytechnic School in Berlin and he soon moved on to the University of Berlin. Weierstrass's work on Abelian functions is generally considered to be his finest, but he made numerous other contributions to many other areas of mathematics. He was one of the first to make systematic use in analysis of representations of functions by power series. He was a superb and very influential teacher, an excellent fencer, and, unlike many mathematicians, he intensely disliked music. His work in 'arithmetizing' analysis led him into a fierce controversy with the constructivist Leopold Kronecker who thought that Weierstrass's widespread use of nonconstructive proofs and definitions was unsound. It is to Weierstrass together with Augustin Cauchy that modern analysis is indebted for its high standards of rigor. Weierstrass gave the first truly rigorous definitions of such fundamental analytical concepts as limit, continuity, differentiability, and convergence. He also did very important work in investigating the precise conditions under which infinite series converged. Tests for convergence that he devised are still in use.

Weil, André (b. Paris, France, 6.5.1906). French mathematician, who has worked in Indian, France, Brazil, Chicago; and latterly at Princeton University. The father of modern algebraic geometry, both through his own discoveries and by means of the associated Weil conjectures that have fuelled further research over a period of 30 years. In conventional geometry shapes are represented by equations which characterize the coordinates of the points that make them up. The coordinates must be quantities such that equations are meaningful; originally they were always real numbers, which meant that the equations were not always solvable. By the 19c the basic field was the complex numbers in which all polynomials can be solved, but this given a rich structure in which rather too much may be going on. Weil introduced geometries over finite fields, so that the mathematician may choose an appropriate basis for his particular problem. A particular class of such problems will be restrictions of problems not yet solved for the general case, such as the Riemann hypothesis. Other problems may feed back into number theory as results about the distribution and density of classes of numbers. Weil could take a collection of rationals and look at the compact subsets of the product space of all their completions. He developed the theory of integration on topological groups, and could make statements about such number-based structures in terms of volumes. Weil studied at the Ecole Normale Supérieure in Paris, and at the universities of Rome and Göttingen. He has held teaching posts in many countries, including posts at the Aligarh Muslim University in India and at the universities of Strasbourg, Sao Paulo in Brazil, and Chicago. In 1958 he moved to the Institute for Advanced Study at Princeton, where he is now professor emeritus. Weil's mathematical work has centered on number theory, algebraic geometry, and group theory. He proved one of the central results in the theory of algebraic fields. His publications include *Foundations of Algebraic Geometry* (1946). The religious philosopher and mystic, Simone Weil, who died in 1943 was his sister.

Weinberg, Steven (b. May 3, 1933; New York City). American physicist. Weinberg graduated from the Bronx High School of Science in New York in 1950. He went on to major in physics at Cornell University, where he gained his bachelor's degree in 1954. From Cornell, Weinberg went in succession to the University of Copenhagen, Princeton University (gaining his PhD there in 1957), the University of California at Berkeley (1960-69), and the Massachusetts Institute of Technology, where he was professor of physics (1969-73). In 1973 he moved to Harvard as Higgins Professor of Physics. In 1979,

Weinberg received the Nobel Prize for physics together with Abdus Salam and his Bronx and Cornell contemporary Sheldon Glashow, for work that was developed by the three men essentially independently. It concerns the unification of two of the four fundamental forces of physics: the electromagnetic interaction (operating between charged particles) and the weak interaction (which accounts for certain radioactive decay processes). The theory, which gives a single explanation of the two (the *Weinberg—Salam theory,* later extended by Glashow), is an example of a guage theory—i.e.a theory changing the scale from one reference frame to another. Electromagnetic and weak interactions are markedly different in their scale, the electromagnetic interaction being 10^{10} times stronger than the weak interaction. The formulation of a single theory encompassing the two was a major triumph for Weinberg, Salam, and Glashow—comparable with Clerk Maxwell's unification of the electric and magnetic field into the electromagnetic field in the 19th century. It went part of the way toward the elusive 'unified fields theory'—a single theory of strong, electromagnetic, weak, and gravitational interactions sought by Albert Einstein, Erwin Schrödinger, and many others. A notable prediction of the theory was the existence of what are called 'neutral currents'—reactions between elementary particles in which no electric charge is exchanged. The main support for the theory came from the experimental observation of neutral currents in 1973. Weinberg has also applied his knowledge of elementary particles to problems of astrophysics and cosmology. He is the author of *Gravitation and Cosmology* (1972) and the more popular *The First Three Minutes* (1977) on the processes of creation of the universe. He is also interested in medieval history.

Weyl, Hermann (b. Elmshorn, Schleswing-Holstein, Germany, 9.11.1885; naturalized American citizen; d. Zürich. Switzerland, 8.12.1955). German/American mathematician, especially noted for the establishment of symmetry as an essential tool in applied mathematics and mathematical physics. He worked at the universities of Zürich (1913-30), Göttingen (1930-3) and Princeton (1933-55). The configuration space in which the results of possible observations lie will not in general have a well-known geometrical structure; or, especially in cases where the space has very many dimensions, a known structure may be present but unrecognized. One way of characterizing the space is to look at the group of symmetries; that is, of transformations of the space into itself. By analogy, it is useful to describe angular momentum in terms of invariance under rotation, i.e. by symmetry in the momentum space associated with the configuration space; and linear momentum by similar reference to linear displacement. In this way concepts originally defined in Euclidean space may again be recognized in the configuration spaces of quantum physics or general relativity (see *Symmetry,* Princeton, 1952). Mathematically speaking, this work of Weyl's appears as the development of the necessary apparatus by, for instance, studying the global properties of Lie groups (extending results of CARTAN) and finding forms (Weyl chambers, the maximal Torus, etc.) to express these structures in a tractable manner. These results can also be regarded as geometry, and Weyl had earlier (pre-1920) been noted as a geometrician. His *Die Idee der*

Riemannschen Fläche [The idea of the Riemann surface] (1913) gave a complete description of the I-dimension complex Riemannian manifold; this object resembles a 2-dimensional Euclidean space in its degree of complexity. Later he developed the programme of uniquely embedding any compact 2-dimensional Riemannian manifold in Euclidean 3-space, thus providing a standard method of visualizing novel structures within a well-known space. Weyl's earliest work was on spectral theory and analytic number theory. Here he proved classic results about the eaquidistribution of sequences of real numbers modulo 1.

Weyl, Hermann (b. Nov. 9, 1885; Elmshorn, now in West Germany; d. Dec. 8, 1955; Zurich, Switzerland). German mathematician. Weyl studied at Göttingen, where he was one of David Hilbert's most outstanding students. He became a co-worker of Hilbert, who influenced his particular interest and his general outlook on mathematics. Weyl taught at the Federal Institute of Technology in Zurich from 1913, and this too had a decisive influence in directing his mathematical interests through the presence there of Albert Einstein. In 1930 he returned to Göttingen to take up the chair vacated by Hilbert. With the Nazis' rise to power in 1933, Weyl, along with many other members of the Göttingen scientific community went into exile in America. Weyl found a post at the Institute for Advanced Study in Princeton along with other exiles such as Einstein and Kurt Gödel. Weyl's mathematical interests, like those of Hilbert, were exceptionally wide, ranging from mathematical physics to the foundations of mathematics. He worked on two areas of pure mathematics: group theory, and the theory of Hilbert space and operators, which, although developed for purely mathematical purposes, later turned out to be precisely the mathematical framework needed for the revolutionary physical ideas of quantum mechanics. Weyl also wrote a number of books on the theory of groups, and he was particularly interested in symmetry and its relation to group theory. One of his most important results in group theory was a key theorem about the application of representations to Lie algebras. Weyl's work on Hilbert space had grown out of his interest in Hilbert's work on integral equations and operators. The theory of Hilbert space (infinite-dimensional space) was recognized by Erwin Schrödinger and Walter Heisenberg in the mid-1920s as the necessary unifying systematization of their theories of quantum mechanics. Wely's contact with Einstein at Zurich was responsible for an interest in the mathematics of relativity, and especially Riemannian geometry, which plays a central role. Weyl initiated the whole project of trying to generalize Riemannian geometry. He himself worked chiefly on the geometry of affinely connected spaces, but this was only one of many generalizations that resulted from his work. Weyl also did similar work on generalizing and refining the basic concepts of differential geometry. All this work was to be of importance for relativity. Weyl's views on relativity were expounded in his book *Raum-Zeit-Materie* (1919; Space-Time-Matter). Weyl' like his teacher Hilbert, was always interested in the philosophical aspects of mathematics. However, in contrast to Hilbert, his general attitude was similar to that of L.E.J. Brouwer with whom he shared constructivist leanings developed from work in analysis. Weyl expounded his philosophical ideas in another book, *Philosophy of*

Mathematics and Natural Science (1949). Unlike Brouwer, however, Weyl was less rigorous in avoiding nonconstructive mathematics, and doubtless his interest in physics contributed to this.

Wheeler, John Archibald (b. July 9, 1911; Jacksonville, Florida). American theoretical physicist. Wheeler was educated at Johns Hopkins University, where he obtained his PhD in physics in 1933. After spending the period 1933-35 in Copenhagen working with Niels Bohr, he returned to America to take up a teaching position at the University of North Carolina. In 1938 he went to Princeton, where he served as professor of physics from 1947 until his move to the University of Texas in 1976 to become professor of physics. wheeler has been active in theoretical physics. One of the problems tackled by him has been the search for a unified field theory. His earlier papers on the subject were collected in 1962 in his *Geometrodynamics*. It was here that he introduced the geon (gravitational— electromagnetic entity), with which he aimed to achieve the unification of the two fields. He also collaborated with Richard Feynman in two paper in 1945 and 1949 on the important concept of action at a distance. They formulated a problem that arises when it is accepted that such action cannot take place instantaneously. If X and Y are at rest and one light-minute apart, then any electromagnetic signal emitted by X will reach Y one minute later. This is described by saying X acts on Y by a retarded effect. But by Newton's third law, to each action there corresponds an opposite and equal reaction. This must mean that from Y to X there should also be an advanced effect acting backward in time. Feynman and Wheeler demonstrated how the advance wave could be eliminated from the model to account for the fact that the universe displays only retarded effects. Wheeler, also made important contributions to nuclear physics. With Niels Bohr he put forward an explanation of the mechanism of nuclear fission. He joined the Los Alamos group exploring the possibility of producing an explosive device using heavy hydrogen in 1949-50.

Whiston, William (b. Dec. 9, 1667; Norton, England; d. Aug. 22, 1752; London, England). British mathematician and geologist. Whiston, the son of a parish priest, was educated at Cambridge University, where he came to the attention of Isaac Newton. He was selected, on Newton's recommendation, to succeed his as Lucasian Professor of Mathematics (1703) and to edit his *Arithmetica universalis* (1707; Universal Arithmetic). In 1710 he was dismissed from the university for his unorthodox religious belief in the Arian heresy, after which time he supported himself by giving public lectures on popular science. Whiston's chief scientific work, *A New Theory of the Earth* (1696), was praised by Newton and by John Locke. In this Whiston followed the tradition recently established by Thomas Burnet in attempting to explain biblical events, such as the Creation, scientifically. The Flood, he believed, was caused by a comet passing close to the Earth on 28 November, 2349 BC. This put stress on the Earth's crust causing it to crack and allow the water to escape and flood the Earth.

Whitehead, Alfred North (b. Feb. 15, 1816; Ramsgate, England; d. Dec. 30, 1947; Cambridge, Massachusetts). British mathematician and philosopher. Whitehead obtained his PhD from Cambridge University, England, in 1884. For the next few years he taught there and met Bertrand Russell (q.v), who was one of his students and with whom he was later to collaborate. Whitehead was one of a growing section of philosophers of science who criticized the deterministic and materialistic views prevalent in 19th-century science. The main theme of this critique was that scientific theories were patterns derived from out way of measuring and perceiving the world and not innate properties of the underlying reality. While in Cambridge his mathematical work reflected this viewpoint, developed in his book *Treatise of Universal Algebra* (1898), which treated algebraic structures as objects worthy of study in their own right, independent of their relationship to real quantities. In 1910 he published, with Russell, the first volume of the vast *Principia Mathematica*, which was an attempt, inspired by the work of Gottlob Frege and Guiseppe Peano, to clarify the conceptual foundations of mathematics using the formal methods of symbolic logic. Whitehead then moved to London, where he taught at University College and later became professor at Imperial College. Here he developed his action-at-a-distance theory of relativity, which challenged Einstein's field-theoretic viewpoint but never gained wide acceptance. In 1924 he emigrated to America and worked at Harvard developing his antimechanistic philosophy of science and a system of metaphysics, until he retired in 1937.

Whittaker, Sir Edmund Taylor (b. Oct. 24, 1873; Birkdale, England; d. Mar. 24, 1956; Edinburgh). British mathematician and physicist. Whittaker studied at Cambridge, and taught there from 1896. In 1906 he became Astronomer Royal for Ireland. From 1912 until 1946 he was professor of mathematics at Edinburgh. Whittaker was primarily a mathematical physicist, and was appropriately much interested in the theory of differential equations. In this field he obtained the general solution of Laplace's equation in three dimensions. This led Whittaker to finding the general integral representation of any harmonic function, which greatly simplified potential theory as well as opening up new areas of research. He also did notable work on the theory of automorphic functions. One of Whittaker's other great interests was in relativity theory. He wrote a number of influential textbooks on analysis, on quantum mechanics and on dynamics, including his *History of the Theories of Aether and Electricity* (1910; revised 1941 and 1951), a comprehensive history of theories of electromagnetism and atomism. Whittaker was known as a superb and highly effective teacher. He was also a deeply religious man with an interest in the philosophical and religious importance of contemporary physics and wrote a number of books on these questions.

Wieland, Heinrich Otto (b. June 4, 1877; Pforzheim, now in West Germany; d. Aug. 5, 1957; Munich, West Germany). German chemist. Wieland was the son of a chemist in a gold and silver refinery. He was educated at the University of Munich where he obtained his PhD in 1901. After teaching at the Munich Technical Institute and the University of Freiburg, Wieland succeeded Richard Willstätter in 1925 as professor of chemistry at the University of Munich, a post he retained until his retirement in 1950. In 1912 Wieland began work on the bile acids. These secretions of the liver had been known for the best part of a century to consist of a large number of substances. He began by investigating three of them: cholic acid, deoxycholic acid, and lithocholic acid, finding that they were all steroids, very similar to each other, and all convertible into cholanic acid. As Adolf Windhaus had derived cholanic acid from cholesterol, an important biological sterol, this led Wieland to a proposed structure for cholesterol. For his contributions to steroid chemistry Wieland was awarded the 1927 Nobel Chemistry Prize. After 1921 Wieland worked on a number of curious alkaloids including toxiferin, the active ingredient in curare, bufotalin, the venom from toads, and phalloidine and amatine, the poisonous ingredients in the deadly amanita mushroom.

Wiener, Norbert (b. Nov. 26, 1894; Columbia, Missouri; d. Mar. 18, 1964; Stockholm). American mathematician. Wiener was a child prodigy in mathematics who sustained his early promise to become a mathematician of great originality and creativity. He is probably one of the most outstanding mathematicians to have been born in the USA. Such was Wiener's precocity that he took his degree in mathematics, from Tufts University, at the age of 14 in 1909. Throughout his life Wiener had many extramathematical interests, especially in biology and philosophy. At Harvard his studies in philosophy led him to an interest in mathematical logic and this was the subject of his doctoral thesis, which he completed at the age of 18. Wiener went from Harvard to Europe to pursue his interest in mathematical logic with Bertrand Russell in Cambridge and with David Hilbert in Göttingen.

After he returned from Europe, Wiener's mathematical interests broadened out but, surprisingly, he was unable to get a suitable post as a professional mathematician and for a time tried such unlikely occupations as journalism and even writing entries for an encyclopedia. In 1919 Wiener finally got a post in the mathematics department of the Massachusetts Institute of Technology where he remained for the rest of his career. After his arrival at MIT Wiener began his extremely important work on the theory of stochastic (random) processes and Brownian motion. Among his other very wide mathematical interests at this time was the generalization of Fourier's work on resolving functions into series of periodic functions (this is known as harmonic analysis). He also worked on the theory of Fourier transforms. During World War II Wiener devoted his mathematical talents to working for the military—in particular to the problem of giving a mathematical solution to the problem of aiming a gun at a moving target. In the course of this work Wiener discovered the theory of the prediction of stationary time series and brought essentially statistical methods to bear on the mathematical analysis of control and communication engineering. From here it was a short step to his important work in the mathematical analysis of mechanical and biological systems, their information flow, and the analogies between them — the subject he named 'cybernetics'. It allowed full rein to his wide interests in the sciences and philosophy and Wiener spent much time popularzing the subject and explaining its possible social and philosophical applications. Wiener also worked on a wide range of other methematical topics, particularly important being his work on quantum mechanics.

Wilkins, John (b. 1614; Fawsley, England; d. Nov. 16, 1672; London). English mathematician and scientist. Wilkins was educated at Oxford University, graduating in 1631. He was a parliamentarian during the English Civil War and become warden of Wadham College, Oxford University. In 1659 he was appointed master of Trinity College, Cambridge University. After the Restoration he lost his post but regained favor to become bishop of Chester. Wilkins' chief contribution to the development of science was his part in founding the Royal Society. His influence can be traced back to his student days at Oxford when he collected around him a lively group of philosophers and scientists who later became founder members of the society in 1662. His own writings covered a wide range of fields and although he had a certain amount of mathematical knowledge he was more a practical scientist. *His Discovery of a World in the Moon* (1638) is a fantasy in which he speculated about the structure of the Moon. A later semimathematical work *Mathematical Magick* deals with the principles of machine design and in it Wilkins argued that perpetual motion is a theoretical possibility. One nonscientific interest to which Wilkins devoted much time was his project of devising a universal language.

Wilkinson, Geoffrey (b. July 14, 1921; Todmorden, England). British inorganic chemist. Wilkinson was educated at Imperial College, London, where, after spending the war years working in North America on the development of the atomic bomb, he finally obtained his PhD in 1946. He later worked at the Massachusetts Institute of Technology and at Harvard, before being elected to the chair of inorganic chemistry at Imperial College.

Wilkinson is noted for his studies of inorganic complexes. He shared the Nobel Prize for chemistry in 1973 with Ernst Fischer (q.v.) for work on 'sandwich compounds'. A theme of Wilkinson's work in the 1960s was the study and use of complexes containing a metal-hydrogen bond. Thus, complexes of rhodium with triphenyl phosphine ($(C_6H_5)_3P$) can react with molecular hydrogen. The compound $RhCI(P(C_6H_5)_3)$, known as *Wilkinson's catalyst,* was the first such complex to be used as a homogeneous catalyst for adding hydrogen to the double bonds of alkenes (hydrogenation). This type of compound can also be used as a catalyst for the reaction of hydrogen and carbon monoxide with alkenes (hydroformulation). It is the basis of industrial low-pressure processes for making aldehydes from ethane and propene.

William Harkness. William Harkness was born in Ecclefechan, Scotland, on December 17, 1837, and when only a year and a half old was taken by his parents to the United States. He was graduated from Rochester Univeristy in 1856 and he received his degree from the New York Homeopathic Medical College in 1862. On August, 1, 1862, he was appointed an aid on the staff at the United States Naval Observatory and on August 24, 1863, professor of mathematics in the United States Navy with the relative rank of lieutenant-commander. After the war, he was attached to the United States monitor *Monadnock* to observe the behavior of her compasses under the influence of the heavy iron armor of the vessel. During the cruise, which extended from Philadelphia around Cape Horn to San Francisco, he completely determined the teerestrial magnetic declination, inclination, and horizontal force at all the principal ports visited. From this time on he traveled extensively, visiting observatories in all parts of the world. On August 7, 1869, he observed the total solar eclipse at Des Monies, Iowa, and at that time discovered the now well-known Coronal line K 1474. He also observed the total solar eclipse in Syracuse, Sicily, in December, 1870. He was one of the original members of the Transit of Venus Commission. Harkness had a highly developed capacity for mechanical invention, and devised many instruments and pieces of apparatus noteworthy for their ingenuity and effectiveness. For many years after his death there was hardly a piece of apparatus in use at the Observatory which, if not actually his work, did not embody essential features which he had suggested. Prior to his time, photographic plates had little of their present importance, since it was impossible to measure on them the relative positions of the centers of the heavenly bodies. Eminent German and English astronomers had tried for years to remedy this defect. Since they could not do so they planned to discard astronomical photography entirely, when Harkness invented the required instrument. He also invented the spherometer caliper, probably the most accurate instrument known for determining the figures of the pivots of astronomical instruments. In 1879 he discovered the theory of the focal curve of achromatic telescopes, now universally used for defining exactly the color corrections. His photographic observations were some of the most notable of his time and are authoritative today. Perhaps he will be best remembered for his work on the solar parallax and its related constants. In this he tried to assign a relative degree of accuracy to observations differing in character, in principle, and in design, and to deduce from the multifarious evidence a precise value of the solar parallax, in which each of the different processes contributes its just share to the final result. While he could not complete such an enormous task, the extent and completeness of the inquiry make it a very valuable historical record. In 1894 Professor Harkness was appointed astronomical director of the Naval Observatory, with complete control of all its astronomical work. He was also made director of the *Nautical Almanac* in June, 1897. He held both of these office until his detachment from duty on December 15, 1899, preliminary to his retirement for age on December 17, 1899, when he was promoted to the rank of rear-admiral. He died of Bright's disease in 1903.

William Huggins. Sir William Huggins, pioneer in the science of astrophysics, was born in London on February 7, 1824. He was educated for two years at the City of London School, and from 1839 by private tutors at home. His education was of the usual literary kind, but his own inclination was always towards science. His earliest fancy was the microscope, and from 1852 he was a member of the Royal Microscopic Society. While he was still a young man, after a brief commercial career, Sir William Huggins succeeded to a private fortune, and thenceforward his sold interest was astronomy. As early as 1854 he was fully equipped with an observatory of his own, fitted with an equatorial telescope of five-inch aperture and other instruments. This observatory was built in his garden at Tulse Hill, in London, where he lived and worked until his death. After becoming an expert in the methods of astronomical observation which were familiar at the time, Sir William was attracted, in

1862, by the idea of spectroscopy, owing to a treatise by Kirchhoff which showed how the dark lines of the solar spectrum could be made to reveal the physical conditions of the sun. Greatly interested in the new vista of knowledge thus opened up, Sir William Huggins set to work at one to apply the same method to the profoundly suggestive investigation of more distant celestial objects. Together with Sir William Allen Miller, he adapted the spectroscope to the study of the stars; and with their new instrument they examined the physical constitution of Betelgeux and Aldebaran, and after these of Sirius. Sir William was the pioneer in this work of searching investigation of the spectra of individual stars, which has revealed entire new worlds of knowledge to the astronomer. For long he worked almost alone in this field. Donati's experiments, though earlier in date, were unsuccessful because his apparatus was unsuitable to the work. Secchi, who was working at stellar spectroscopy at about the same time as Huggins, made no detailed examination of special stars, but swept rapidly through the heavens in search of general results and was thus the first to classify the stars according to their spectra. The careful analytic methods of Huggins, which were quite new to astronomy, resulted in establishing a general similarity, and many profoundly interesting differences, between the sun and the other stars. The next spectroscopic discovery by Sir William Huggins was of epoch-making importance. He was examining, in August, 1864, a planetary nebula in Draco, and found that it gave a "bright-line" spectrum, thus proving its gaseous nature. Eight other nebulae which he then examined gave similar results; and the 1868 the Orion nebula was found to yield a spectrum which showed its gaseous constitution, and that it consisted largely of hydrogen, a fact which confirmed the latest belief of Sir William Hershcel. The Andromeda nebula, on the contrary, gave a continuous spectrum. These discoveries were of immense importance in clearing away the difficulties in the way of the nebular hypothesis. Several new stars now came under the test of the spectroscope in the hands of Sir William Huggins. In 1886 he examined the "nova" in Corona Borealis, and in 1891 the brilliant new star in Auriga. He found that these sudden visitants were distinguished from ordinary stars by the presence of certain extremely significant bright lines in their spectra, showing evidence of a conflagration of hydrogen on a most stupendous scale. On this evidence he based his well-known explanation of "novae". He says, "The case is that of the casual near approach of two bodies previously possessing considerable velocities in space, such a near approach being far less improbable than an actual or partial collision. Enormous disturbances of a tidal nature would inevitably follow such approach, and produce sufficiently great changes of pressure in the interior of the bodies to cause tremendous eruptions from within, similar in kind to solar outbursts, but immensely greater." The most important of all Sir William's discoveries was, however, his adoption of spectroscopic methods for determining stellar motions in the line of sight. The method resulted immediately in most momentous results Sirius first, and many other brilliant stars afterwards, were found to have extremely rapid movements either towards or away from the earth. Many years will pass before the uses of this method have been exhausted. Sir William was also a pioneer in the photography of stellar spectra. After several vain attempts, he succeeded, in 1879, in obtaining photographs of several of these; and in 1882 photographed the solar coronal rays. By all these indefatigable researches, Sir William Huggins ranks among the great astronomers; and he was undoubtedly the founder of the modern science of astrophysics. He was honored with all kinds of distinctions at home and abroad. In 1900 he was president of the Royal Society, and for many years president of the Astronomical Society. He was one of twelve chosen as the first members of the Order of Merit in 1902. He died in London on March 12, 1910. In 1875 Sir William Huggins married Margaret Murray, who was already a keen student of the heavens; she became his sole assistant, and all his papers after their marriage were signed by both. In 1901 Lady Huggins was elected honorary members of the Royal Astronomical Society.

Williams, Robert R. (b. Feb. 16, 1886; Nellore, India; d. Oct 2, 1965; Summit, New Jersey). American chemist. Williams, a Baptist missionary's son, was educated at the universities of Ottawa, Kansas, and Chicago. He began his career in government service, serving as chemist to the Bureau of Science in Manila before returning to America where he worked at the Bureau of Chemistry in the agriculture department until 1918. He then moved into industry, working first for Western Electric before moving to the Bell Telephone Laboratories in 1924, where he directed the chemistry laboratory until 1945. With considerable single-mindedness Williams, early in his career, set himself the task of isolating the cause of beri-beri. As early as 1896 Christian Eijkman had shown that it was a deficiency disease while Casimir Funk had demonstrated that the vitamin whose absence caused the disease was an amine. Beyond that nothing was known when Williams began his work in the Philippines. Working mainly in his spare time, in 1934 he managed to isolate, from several tons of rice husks, enough of the vitamin B_1 to work out its formula. In 1937 he succeeded in synthesizing it. His brother, Roger, also a chemist, discovered pantothenic acid, another important vitamin in the B complex.

Williamson, Alexander William (b. May 1, 1824; London; d. May 6, 1904; Haslemere, England). British chemist. Williamson's father was a clerk in the East India Company. After his retirement in 1840 the family lived on the Continent, where Williamson was educated. He studied at Heidelberg and at Giessen (under Justus von Liebig), from where he received his PhD in 1846. He also studied mathematics in Paris. In 1849 he was appointed to the chair of chemistry at London University, a post he occupied until 1887. Between 1850 and 1856 Williamson showed that alcohol and ether both belong to the water type. Type theory, developed by Charles Gerhardt and Auguste Laurent, was based on the idea that organic compounds are produced by replacing one or more hydrogen atoms of inorganic compounds (which form the types) by radicals. Using the correct formula for alcohol (which he had recently established) Williamson represented the water type as: H_2O (water); C_2H_5OH (alcohol); $C_2H_5OC_2H_5$ (ether), where the H of water is progressively replaced by C_2H_5. A further contribution to chemical theory was his demonstration (in 1850) of reversible reactions: two substances, A and B, react to form the products X and Y, which in turn react to produce the original A and B. Under certain conditions the system could be in dynamic equilibrium, when the amount of A and B reacting to form X and Y is equal to the amount of A and B produced by X and Y. He is remembered for what is now known as *Williamson's synthesis,* a method of making ethers by reacting a sodium alcoholate with a haloalkane.

William Thomson, Lord Kelvin. William Thomson, like many other British physicists, was an extremely precocious child. He was born in Belfast in 1824, the son of a prominent professor of mathematics. One story is told that when only eight years old he used to attend his father's university lectures with great enjoyment. At eleven he matriculated at the University of Glasgow and was graduated second in his class. He wrote his first original paper in mathematics before his twentieth birthday. At the age of seventeen he went to Cambridge, where he took an advanced degree four years later. He then went to Paris to study with the French chemist and physicist Henri Victor Regnault, a pioneer experimenter in what eventually became the science of physical chemistry. At one point in his career Thomson became famous for his calculation of the age of the earth, which he placed at between 20 million and 400 million years. This faulty figure, based on a mistaken concept, had the serendipitous result of obliging evolutionists to evolve the mutation theory to explain how such great evolutionary changes as have been observed could take place in so short a time. Thomson was influential in getting acceptance in the scientific world for theories devised by Joule and Faraday. He also calculated the value of absolute zero as—273°C (the accepted figure today is—273.15°C); in the SI system of measurements temperatures are measured in units called kelvins in honor of Thomson's work in the field. (Kelvins are equal to Celsius degrees; the system of expressing temperatures as ^{o}K has been abandoned, by the way, and 0°C is now 273K, without the degree sign.) Thomson's work on the theory of heat sprang in part from his dissatisfaction with Carnot's conclusion that heat is a substance. Carnot's theory had originated with a mistaken analogy between heat and his column-of-water device in which "a quantity of water falling through a fixed distance from a given height produces an invariable quantity of motive power, the water being fully transferred from its original height to a reservoir at a lower level." Thomson rightly suspected that a given quantity of heat could not be retrieved in the same way as water lying in a lower reservoir and that some heat would always be lost for good during such a transfer. Carnot's error provided fertile ground for later considerations by Thomson and others on the impossibility of perpetual motion. Although Thomson had begun his career with the discovery of a simple analogy between the flow of heat from a hot body and the flow of an electric field from an electrified body, he never fully accepted the physical existence of the "electric fluid" nor of discrete atoms. He devoted much of his studies to ridding physics of conceptions such as these which he called "physical hypotheses." Under the conviction that a continuous fluid is less of a physical hypothesis than a system of atoms, he returned to the Cartesian doctrine of vortices and was the first to formulate a mathematical model of a Universe of vortices. Although the theory did not work, it was greatly admired for its mathematical beauty. Thomson's ability to solve problems of the propagation of waves of all kinds was invaluable in the laying of the first transatlantic cable, which employed a signaling device of his invention. Thomson's dislike for what he called physical hypotheses was particularly useful for advancing the study of thermodynamics, as it caused him to eliminate the caloric concept from Carnot's theory. He was aware of the apparent contradiction between the two thermodynamic ideas: Heat is not a conserved fluid but is a form of motion. Yet is the fall of heat from a higher temperature to a lower temperature that makes it possible to perform work. Thomson's accomplishment was to accept both ideas as ture, embodying them in the first and second laws of thermodynamics. This work independently paralleled the efforts of Clausisu, although Clausius used engines in his theory while Thomson preferred refrigerators, which are engines run backwards. In 1882 Queen Victoria raised Thomson to the peerage, and he became Lord Kelvin of Largs. His name was taken from the Kelvin River near Glasgow. He was the first and last Lord Kelvin, for he died childless. He was entombed in Westminster Abbey next to Sir Isaac Newton.

Winograd, Terry Allen (b. Takoma Park, Md, USA, 24.2.1946). American computer scientist and psychologist. He graduated in linguistics from University College London, and then studied for a PhD (1970) in applied mathematics at MIT. Since 1972 he has been a research consultant for Rank-Xerox and since 1974 has held a professorial post in computer science and linguistics at Stanford University. Winograd is one of the leading advocates of the use of artificial intelligence in the development of psychological theories. His particular concern has been the development of computer programs that model the brain's ability to comprehend natural language. In *Understanding Natural Language* (NY, 1972) Winograd described an innovative program which was the first to take account of the relationship between language, thought and action. Since then he has been concerned with computer models of semantic and syntactic aspects of comprehension and with examining the implications of the procedural/declarative distinction (i.e. 'knowing how' versus 'knowing that') for theories of knowledge representation.